AF358734

Fracture and Fatigue Assessments of Structural Components

Fracture and Fatigue Assessments of Structural Components

Editor

Alberto Campagnolo

MDPI • Basel • Beijing • Wuhan • Barcelona • Belgrade • Manchester • Tokyo • Cluj • Tianjin

Editor
Alberto Campagnolo
University of Padova
Italy

Editorial Office
MDPI
St. Alban-Anlage 66
4052 Basel, Switzerland

This is a reprint of articles from the Special Issue published online in the open access journal *Applied Sciences* (ISSN 2076-3417) (available at: https://www.mdpi.com/journal/applsci/special_issues/Structural_Components).

For citation purposes, cite each article independently as indicated on the article page online and as indicated below:

LastName, A.A.; LastName, B.B.; LastName, C.C. Article Title. *Journal Name* **Year**, *Volume Number*, Page Range.

ISBN 978-3-03943-729-0 (Hbk)
ISBN 978-3-03943-730-6 (PDF)

© 2020 by the authors. Articles in this book are Open Access and distributed under the Creative Commons Attribution (CC BY) license, which allows users to download, copy and build upon published articles, as long as the author and publisher are properly credited, which ensures maximum dissemination and a wider impact of our publications.
The book as a whole is distributed by MDPI under the terms and conditions of the Creative Commons license CC BY-NC-ND.

Contents

About the Editor

Alberto Campagnolo was born on February 27th, 1987. In 2009, he obtained his bachelor's degree cum laude in Mechanical Engineering from the University of Padova with the score 110/110. In 2012, he obtained his master's degree cum laude in Mechanical Engineering from the University of Padova with the score 110/110. In 2016, he obtained his PhD in Mechatronics and Product Innovation Engineering from the University of Padova with the dissertation "Local Approaches Applied to Fracture and Fatigue Problems" (Supervisors: Prof. Paolo Lazzarin and Prof. Filippo Berto). In 2016, he won a two-year junior research grant with the title "Development, experimental validation and implementation in commercial FE codes of methods for the prediction of the structural integrity of welded structures subjected to multiaxial cyclic loadings", at the Department of Industrial Engineering of the University of Padova. In 2018, he became Assistant Professor of Machine Design at the Department of Industrial Engineering of the University of Padova, and in the same year, he obtained the national abilitation for Associate Professor in Machine Design. He is author of around 100 scientific publications, 50 of which were published in international journals with impact factors, while 50 were published in the proceedings of international or national conferences. He has served on the editorial board of the international journals *Applied Sciences* (Basel, MDPI) and *Mathematical Problems in Engineering* since 2018. His research deals with the development of local approaches for structural durability analysis of welded components and structures, the analysis of the fatigue behavior of notched components in metallic materials, and adoption and development of the electrical potential drop method for monitoring fatigue crack initiation and propagation. Within these fields, he has international collaborations with several researchers, among them Prof. Keisuke Tanaka (Meijo University), Prof. Majid R. Ayatollahi (Iran University of Science and Technology), Prof. Michael Vormwald (Technische Universität Darmstadt), and Dr. Jurgen Bar (Universität der Bundeswehr).

Editorial

Special Issue on Fracture and Fatigue Assessments of Structural Components

Alberto Campagnolo

Department of Industrial Engineering, University of Padova, Via Venezia 1, 35131 Padova, Italy; alberto.campagnolo@unipd.it; Tel.: +39-049-827-7475

Received: 4 August 2020; Accepted: 24 August 2020; Published: 11 September 2020

Abstract: This Special Issue covers the broad topic of structural integrity of components subjected to either static or fatigue loading conditions, and it is concerned with the modelling, assessment and reliability of components of any scale. Dealing with fracture and fatigue assessments of structural elements, different approaches are available in the literature. They are usually divided into three subgroups: stress-based, strain-based and energy-based criteria. Typical applications include materials exhibiting either linear-elastic or elasto-plastic behaviours, and plain and notched or cracked components subjected to static or cyclic loading conditions. In particular, the articles contained in this issue concentrate on the mechanics of fracture and fatigue in relation to structural elements from nano- to full-scale and on the applications of advanced approaches for fracture and fatigue life predictions under complex geometries or loading conditions.

Keywords: fracture; fatigue; notch; crack; metal; structure; welded joint; FEM

1. Introduction

This Special Issue was introduced to collect the latest research on fracture and fatigue of structural elements, and more importantly, to address present challenging issues in the context of the integrity of structures from nano- to full-scale and components under complex loading conditions. In light of the above, this Special Issue embraces interdisciplinary works aimed at understanding and deploying physics of fatigue and failure phenomena, advanced experimental and theoretical failure analysis, modelling of the structural response with respect to both local and global failures, and providing structural design approaches to prevent engineering failures. Original contributions from engineers, mechanical and material scientists, computer scientists, physicists, chemists, and mathematicians are presented, following both experimental, numerical and theoretical approaches. There were 34 papers submitted to this Special Issue, and 11 papers were accepted (i.e., 33% acceptance rate).

2. Fracture

A number of papers in this Special Issue are specifically devoted to fracture mechanics problems [1–3]. Different approaches have been adopted, including experimental investigations, theoretical models, and numerical simulations.

Gallo and Sapora [1] have proposed a method which could be useful for predicting the static failure of micro- and nano-electromechanical systems (MEMS, NEMS). In more detail, the authors have applied a coupled stress-energy approach—so-called Finite Fracture Mechanics—to predict the failure load of notched nano-components made of single crystal silicon.

In [2], the authors developed a two-dimensional ordinary state-based peridynamic modeling of mode-I delamination growth in a double cantilever composite beam test using revised energy-based failure criteria. The proposed analytical model has successfully been validated against experimental results.

Finally, in [3], a stress field analytical model of the wellbore coal rock has been established by considering the irregularity of the cleat distribution and the influence of the cleat filler. The analytical model has been compared with numerical simulations obtaining a good agreement.

3. Fatigue

In this Special Issue, the fatigue phenomenon has been investigated from experimental, numerical, and theoretical points-of-view in different papers [4–11]. Contributions [4–8] were focused on the application of advanced analytical or numerical approaches to predict the experimental fatigue strength of laboratory specimens; while, on the other hand, papers [9–11] were mainly devoted to the fatigue assessment of full-scale, real structures undergoing complex loading conditions.

Ronchei and co-authors [4], have applied a critical plane-based multiaxial fatigue criterion for the fatigue life assessment of Ti-6Al-4V notched specimens. The accuracy of the proposed criterion has successfully been evaluated through experimental data available in the literature.

In [5], the authors proposed an energy-based approach for the fatigue life estimation of welded joints through thermal-graphic measurement. A model based on intrinsic energy dissipation was applied to high strength steel welded joints showing a good agreement between estimations and experimental results.

Paper [6] was devoted to presenting a multiscale fatigue damage evolution model for describing both the mesoscopic voids propagation and fatigue damage evolution process, reflecting the progressive degradation of metal components in the macro-scale. A method of defect classification was employed to implement 3D reconstruction technology based on the micro-computed tomography scanning damage data with FE simulations. The predictions were validated through a comparison with experimental data.

The aim of [7] was to characterize the propagation of fatigue cracks using the damage index derived by various acoustic features of ultrasonic guided waves. The method has been validated by monitoring the fatigue crack propagation in a steel plate-like structure.

In [8], the authors have formulated a numerical model to simulate the thorough failure process on concrete, ranging from microcracks growth, crack coalescence, macrocrack formation and propagation, to the final rupture. The model has been applied to simulate the fatigue rupture of three-point bending concrete beams, observing a good agreement between numerical results and experimental observations available in literature.

Paper [9] focused on the assessment of fatigue life and characterization of the fatigue crack behavior of an aluminum scroll compressor, taking into account both mean stress effects and elastic-plastic behavior of the material. The authors took advantage of both analytical and numerical models.

In [10], the authors have investigated the factors inducing fatigue crack initiation from the positioning block weld toe of metro bogie frame, which is the critical safety part of the urban metro vehicle. Metallographic analyses were employed to study the failure modes and fracture characteristics of the weld toe of positioning block. On-track testing was carried out to obtain acceleration and the stress response information of the bogie, and to investigate which factors could be optimized in order to reduce the failure probability.

Finally, the aim of [11] was to develop a master–slave model with fluid-thermo-structure interaction for the thermal fatigue life prediction of a thermal barrier coat in a nozzle guide vane. The master–slave model integrates the phenomenological life model, multilinear kinematic hardening model, fully coupling thermal-elastic element model, and volume element intersection mapping algorithm to improve the prediction precision of thermal fatigue life.

Funding: This research received no external funding.

Acknowledgments: This Special Issue would not have been made possible without the irreplaceable contributions of valuable authors coming from many different countries, hardworking and professional reviewers, and dedicated editorial team of Applied Sciences, an international, peer-reviewed journal that is free for readers embracing all aspects of applied sciences. I would like to take this opportunity to record my sincere gratefulness to all reviewers.

Finally, I place on record my gratitude to the editorial team of Applied Sciences, and special thanks to Daria Shi, Managing Editor, from MDPI Branch Office, Beijing.

Conflicts of Interest: The author declares no conflict of interest.

References

1. Gallo, P.; Sapora, A. Brittle failure of nanoscale notched silicon cantilevers: A finite fracture mechanics approach. *Appl. Sci.* **2020**, *10*, 1640. [CrossRef]
2. Jiang, X.-W.; Guo, S.; Li, H.; Wang, H. Peridynamic modeling of mode-i delamination growth in double cantilever composite beam test: A two-dimensional modeling using revised energy-based failure criteria. *Appl. Sci.* **2019**, *9*, 656. [CrossRef]
3. Zhao, X.; Wang, J.; Mei, Y. Analytical model of wellbore stability of fractured coal seam considering the effect of cleat filler and analysis of influencing factors. *Appl. Sci.* **2020**, *10*, 1169. [CrossRef]
4. Ronchei, C.; Carpinteri, A.; Vantadori, S. Energy concepts and critical plane for fatigue assessment of Ti-6Al-4V notched specimens. *Appl. Sci.* **2019**, *9*, 2163. [CrossRef]
5. Mi, C.; Li, W.; Xiao, X.; Berto, F. An energy-based approach for fatigue life estimation of welded joints without residual stress through thermal-graphic measurement. *Appl. Sci.* **2019**, *9*, 397. [CrossRef]
6. Bao, Y.; Yang, Y.; Chen, H.; Li, Y.; Shen, J.; Yang, S. Multiscale damage evolution analysis of aluminum alloy based on defect visualization. *Appl. Sci.* **2019**, *9*, 5251. [CrossRef]
7. Jin, H.; Yan, J.; Li, W.; Qing, X. Monitoring of fatigue crack propagation by damage index of ultrasonic guided waves calculated by various acoustic features. *Appl. Sci.* **2019**, *9*, 4254. [CrossRef]
8. Wu, B.; Li, Z.; Tang, K.; Wang, K. Microscopic multiple fatigue crack simulation and macroscopic damage evolution of concrete beam. *Appl. Sci.* **2019**, *9*, 4664. [CrossRef]
9. Park, S.-Y.; Lee, J.; Heo, J.-T.; Lee, G.B.; Kim, H.H.; Choi, B.-H. Assessment of fatigue lifetime and characterization of fatigue crack behavior of aluminium scroll compressor using C-specimen. *Appl. Sci.* **2020**, *10*, 3226. [CrossRef]
10. Wang, W.; Bai, J.; Wu, S.; Zheng, J.; Zhou, P. Experimental investigations on the effects of fatigue crack in urban metro welded bogie frame. *Appl. Sci.* **2020**, *10*, 1537. [CrossRef]
11. Guan, P.; Ai, Y.; Fei, C.; Yao, Y. Thermal fatigue life prediction of thermal barrier coat on nozzle guide vane via master–Slave model. *Appl. Sci.* **2019**, *9*, 4357. [CrossRef]

© 2020 by the author. Licensee MDPI, Basel, Switzerland. This article is an open access article distributed under the terms and conditions of the Creative Commons Attribution (CC BY) license (http://creativecommons.org/licenses/by/4.0/).

Article

Brittle Failure of Nanoscale Notched Silicon Cantilevers: A Finite Fracture Mechanics Approach

Pasquale Gallo [1,*] and Alberto Sapora [2,*]

[1]   Department of Mechanical Engineering, Aalto University, P.O. Box 14100, FIN-00076 Aalto, Finland
[2]   Department of Structural, Building and Geotechnical Engineering, Politecnico di Torino,
      Corso Duca degli Abruzzi 24, 10129 Torino, Italy
*   Correspondence: pasquale.gallo@aalto.fi (P.G.); alberto.sapora@polito.it (A.S.)

Received: 5 February 2020; Accepted: 25 February 2020; Published: 29 February 2020

Featured Application: The work provides an extremely useful method to predict the static failure of Micro- and Nano-Electromechanical Systems (MEMS, NEMS). The Finite Fracture Mechanics approach may have an enormous impact on the failure characterization of notched and cracked components in the field of nanodevices.

Abstract: The present paper focuses on the Finite Fracture Mechanics (FFM) approach and verifies its applicability at the nanoscale. After the presentation of the analytical frame, the approach is verified against experimental data already published in the literature related to in situ fracture tests of blunt V-notched nano-cantilevers made of single crystal silicon, and loaded under mode I. The results show that the apparent generalized stress intensity factors at failure (i.e., the apparent generalized fracture toughness) predicted by the FFM are in good agreement with those obtained experimentally, with a discrepancy varying between 0 and 5%. All the crack advancements are larger than the fracture process zone and therefore the breakdown of continuum-based linear elastic fracture mechanics is not yet reached. The method reveals to be an efficient and effective tool in assessing the brittle failure of notched components at the nanoscale.

Keywords: finite fracture mechanics; nanoscale; silicon; brittle; notch; fracture; nanodevice

---

1. Introduction

Recent technological developments have enabled the fabrication of electronic devices with high-density integration. Small size components, e.g., at the nanometer scale, can be fabricated with different shapes including features such as notches and may have defects such as cracks [1,2]. These circumstances have brought problems commonly addressed by fracture mechanics and fatigue theory to a completely new scale level, raising several new questions, experimental challenges, but also attractive new scientific possibilities [3–5]. The demand for static and fatigue assessment of nanoscale components is increasing, on the one hand, and the validity of continuum-based approaches is questioned on the other. Indeed, at a very small scale, the simplification of a body as continuum and homogeneous may not hold, and the discrete nature of atoms should be considered [6–8]. Clarification of these aspects could not only bring enormous development in the field of nanotechnology, but macroscale could benefit as well, e.g., multi-scale modeling of fatigue with focus on short cracks and interaction with local micro-structure [9,10], atomistic investigation of stresses, strains and grain boundaries [11], fracture properties of advanced materials [12,13], experimental evaluation of fatigue curve of microscale samples [14].

Several studies have recently quantified the small scale breakdown of continuum-based linear elastic fracture mechanics (LEFM) and have demonstrated that continuum-based methods are still valid at small scale if that limit is not reached [15–18]. The breakdown was expressed in terms of the ratio between the crack singular stress field length, which surprisingly is still characterizing the crack at a very small scale [19–21], and the fracture process zone. When this ratio is in the order of 4–5, continuum-based LEFM breaks down and the discrete nature of atoms cannot be ignored. Some researchers have gone at a further small size, and have tried to propose approaches that consider the atomic structures and would be valid at different scales. In this regard, it is worth mentioning those based on Griffith's criterion applied to single crystal silicon [22,23] and graphene [24], and more recently on averaged strain energy density concept [17]. On the other hand, as briefly mentioned earlier, when that limit is not reached, continuum-based approaches should be valid. This conclusion opens up a large number of possibilities in the field of nanotechnology since well-known tools could be extended to small scale for, e.g., nanoscale mechanical characterization, static/fatigue assessment of cracked or notched components [14,25]. For example, the continuum formulation of averaged strain energy density concept has been demonstrated to be valid for nanoscale notched-samples [26,27]. Similarly, by using the Theory of Critical Distances (TCD) [28,29], a fast and extremely simple method was proposed to evaluate with high accuracy the fracture toughness of single crystal silicon by using notched samples [30]. At small scales, indeed, the fabrication of pre-cracks is extremely difficult, since an atomic sharp crack is ideally necessary to obtain a reliable fracture toughness when dealing with brittle materials. Even for small values of the tip radius, the defect would behave as a notch and would result in an apparent fracture toughness [19,31]. In regard to the experimental challenges, it is also worth mentioning a recent study on experimental conditions affecting the measurement of the fracture toughness in elastic-plastic materials [32].

Among well-known fracture models, the Finite Fracture Mechanics (FFM) approach has several similarities with the methods mentioned before [33]. The FFM was originally proposed to deal with crack initiation in brittle material and to overcome the limitation of classical fracture mechanics, which assumes the pre-existence of a crack and deals only with its growth [34,35]. The FFM, instead, is a coupled criterion, i.e., it requires two conditions to be full-filled simultaneously: One based on a stress requirement and the other involving the energy balance. When these conditions are met, there is an instantaneous formation (or growth) of a crack of finite size (finite step). This assumption drastically re-defines the concept of crack growth, which becomes a structural parameter rather than a simple material constant. Over the past decades, the approach has been applied or extended successfully to a large variety of problems, both in static and fatigue, by considering different materials, features and loading conditions, e.g., to rounded V-notches made of ceramic, metallic and plastic materials [36,37], crack at interfaces and at bi-material junctions [38,39], 3-D failure onset from sharp V-notch edge [40], failure initiation at the atomic scale by means of molecular simulations [41], multiaxial loading conditions and notch sensitivity [42,43], moderate and large scale yielding regimes [44]. Furthermore, FFM predictions have been recently proved to be very close to those by the powerful cohesive zone model (CZM) in different research frameworks [45–48], so one can use FFM for preliminary sizing in structural design, letting the CZM for subsequent study refinements. By considering the advantages of the FFM approach and the similitude with other methods successfully applied to small scale specimens, it is worth investigating the validity of the FFM at the nanoscale. FFM would be, indeed, a very useful method for the brittle failure characterization of nanostructures.

By considering experiments available in the literature, the present paper verifies the applicability of the FFM approach at the nanoscale. At first, the analytical frame of FFM is presented; subsequently, recent experimental tests on nanoscale notched cantilevers, made of single crystal silicon, are briefly reviewed and presented; finally, the results are discussed. Only mode I loading condition and brittle fracture are considered. The work represents an additional proof that, when far from the small scale breakdown of continuum-based LEFM, classic concepts such as FFM can be employed successfully to characterize the fracture process of nanodevices.

2. Materials and Methods

2.1. Fundamentals of FFM Approach for Blunt V-Notches

The FFM approach [35] assumes that a crack advances of a finite length l when two criteria, one based on stress and the other on the energy balance, are full-filled simultaneously. The stress criterion requires that the average stress $\sigma_y(x)$ upon the crack advance l is higher than the material tensile strength σ_u:

$$\int_0^l \sigma_y(x)\mathrm{d}x \geq \sigma_u l, \tag{1}$$

where (x, y) is the Cartesian coordinate system centered at the notch root (Figure 1). The energy criterion ensures that the energy available for a crack increment l is higher than the energy necessary to create the new fracture surface. Irwin's relationship allows expressing the requirement in the form

$$\int_0^l K_I^2(c)\mathrm{d}c \geq K_{Ic}^2 l, \tag{2}$$

$K_I(c)$ and K_{Ic} being the stress intensity factor (SIF) related to a crack of length c stemming from the notch root (Figure 1) and the fracture toughness, respectively. At incipient failure, Equations (1) and (2) coalesce into a system of two equations in two unknowns: The critical crack advancement l_c and the failure load.

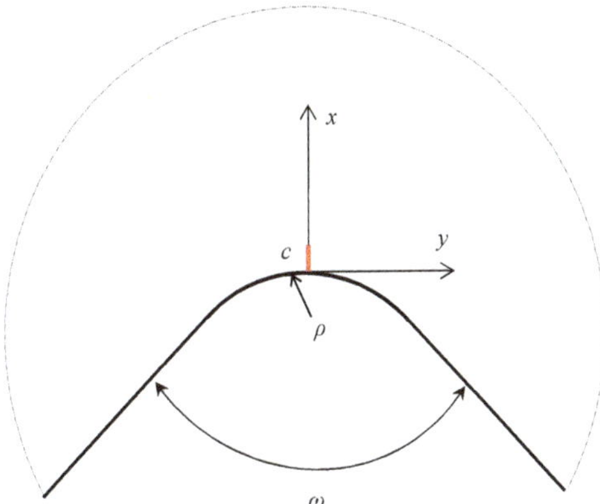

Figure 1. Blunt V-notch geometry: c represent the length of a crack stemming from the notch tip.

Focusing on blunt V-notched geometries [37] and assuming the notch tip radius ρ sufficiently small with respect to the notch depth a (Figure 2), the stress field along the notch bisector can be expressed as [49]:

$$\sigma_y(x) = \frac{K_I^{V,\rho}}{[2\pi(x+r_0)]^{1-\lambda}}\left[1+\left(\frac{r_0}{x+r_0}\right)^{\lambda-\mu}\eta_\theta(0)\right], \tag{3}$$

where $K_I^{V,\rho}$ is the apparent generalized (or notch) SIF, and

$$r_0 = \frac{\pi-\omega}{2\pi-\omega}\rho. \tag{4}$$

The parameters λ, μ and $\eta_\theta(0)$ depend on the notch amplitude ω, and their values are reported in Table 1. Furthermore, supposing a crack of length c sufficiently small with respect to the notch depth a (Figures 1 and 2), the following relationship was proposed for the SIF K_I [50]:

$$K_I(c) = \frac{\beta K_I^{V,\rho} c^{\lambda-0.5}}{\left[1 + \left[\frac{r_0}{c}\left(\frac{\beta}{\alpha}\right)^{\frac{1}{1-\lambda}}\right]^m\right]^{\frac{1-\lambda}{m}}}, \tag{5}$$

where

$$\alpha = 1.12\sqrt{\pi}\,\frac{[1+\eta_\theta(0)]}{(2\pi)^{1-\lambda}}, \tag{6}$$

and β,m are provided in Table 1.

Table 1. Dimensionless parameters used in the present analysis as a function of the notch amplitude ω. See [37] as reference.

ω (deg)	m	λ	μ	$\eta_\theta(0)$	β
0	1.82	0.5000	−0.5000	1.000	1.000
33	1.45	0.5021	−0.4515	1.035	1.005
48	1.38	0.5040	−0.4285	1.007	1.010
59	1.35	0.5075	−0.4105	0.9700	1.017
68	1.34	0.5122	−0.3950	0.9310	1.030
150	1.22	0.7520	−0.1624	0.2882	1.394

Within brittle structural behavior, it is legitimate to suppose that failure takes place as soon as the apparent generalized SIF reaches its critical value $K_I^{V,\rho} = K_{Ic}^{V,\rho}$, usually termed as apparent generalized fracture toughness. As expected, FFM can be implemented by inserting Equations (3) and (5) into Equations (1) and (2), respectively, and integrating. Simple analytical manipulations lead to the following two coupled equations:

$$\frac{K_{Ic}^{V,\rho}}{\sigma_u r_0^{1-\lambda}} = f(\bar{l}_c), \tag{7}$$

and

$$r_0 \frac{\sigma_u^2}{K_{Ic}^2} = \frac{h(\bar{l}_c)}{f^2(\bar{l}_c)}, \tag{8}$$

where the functions f, h rising from the integration procedure are [37]:

$$f(\bar{l}_c) = \frac{\bar{l}_c(2\pi)^{1-\lambda}}{\left[\frac{(\bar{l}_c+1)^\lambda-1}{\lambda}\right] + \eta_0(0)\left[\frac{(\bar{l}_c+1)^\mu-1}{\mu}\right]} \tag{9}$$

$$h(\bar{l}_c) = \frac{\bar{l}_c}{\displaystyle\int_0^{\bar{l}_c} \frac{\bar{c}^{(2\lambda-1)}\beta^2}{\left\{1 + \left[\left(\frac{\beta}{\alpha}\right)^{\frac{1}{1-\lambda}}\frac{1}{\bar{c}}\right]^m\right\}^{\frac{2(1-\lambda)}{m}}}d\bar{c}}, \tag{10}$$

$\bar{l}_c = l_c/r_0$ and $\bar{c} = c/r_0$. The system of Equations (7) and (8) is solved numerically through a simple MATLAB® code: For a given structure, after extracting l_c from Equation (8), the apparent generalized fracture toughness can be estimated through Equation (7).

2.2. Review of Recent In Situ Experimental Tests

The analytical frame presented in Section 2.1 is verified against experimental tests recently published by Gallo et al. [30]. That work was originally proposed to characterize the fracture toughness of silicon by using the TCD. In detail, in situ fracture tests were carried out in a transmission electron microscope (TEM). Four specimens were considered, made of single crystal silicon, and fabricated by a focused ion beam (FIB) processing system. The fabrication process and orientation are simplified in Figure 2 while the detailed procedure is provided in [30]. Geometry and example of a sample are presented in the lower-half of Figure 2, whilst Table 2 gives the geometrical details.

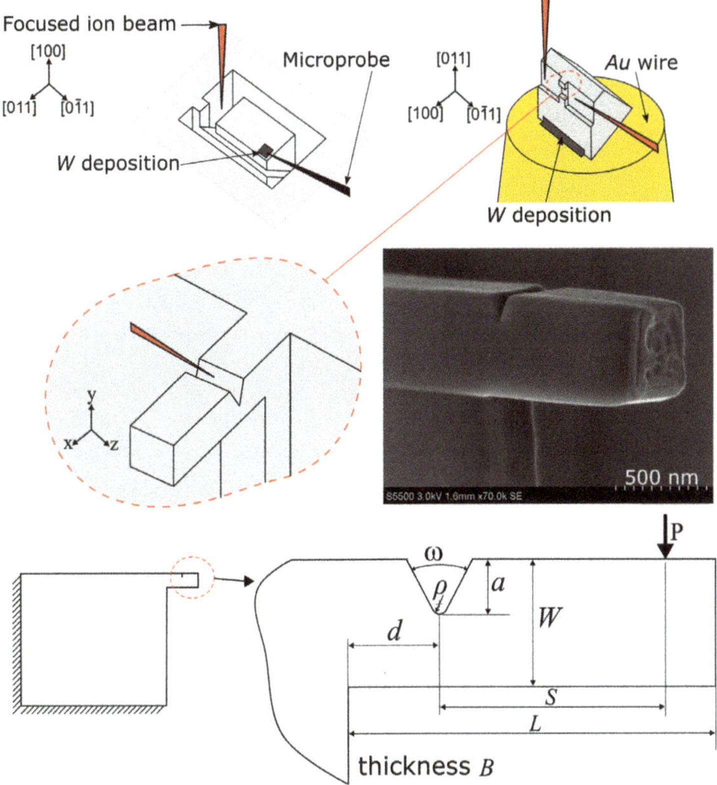

Figure 2. Example of main steps of fabrication process, orientation, final sample (specimen 2) and geometrical details of the blunt V-notched nano-cantilevers. At first, a block is carved from bulk single crystal silicon plate; later, the block is positioned on a gold wire and the nano-cantilever is cut; the notch is finally introduced by focused ion beam (FIB).

Table 2. Geometrical parameters of the blunt V-notched nano-cantilevers (see Figure 2).

Sample	ω (deg)	ρ (nm)	a (nm)	d (nm)	S (nm)	L (nm)	B (nm)	W (nm)	P_f (μN)
1	33	≈10	155	217	837	1307	494	502	45.33
2	48	≈14	161	96	682	1019	529	544	84.80
3	59	≈20	179	119	651	1201	484	495	65.11
4	68	≈6	144	87	875	1252	454	456	30.84

The samples have different notch radii ρ and opening angles ω measured at the best of authors capability by means of scanning electron microscope (SEM). It is worth noting that the samples have an a/W ratio varying between 0.3 and 0.36, while ρ/a varies between 0.07 and 0.11. Experimental mechanical characterization at the considered scale gave an ultimate strength σ_u of 13.9 GPa, and a fracture toughness K_{Ic} of approximately 1 MPa m$^{0.5}$. The latter is determined experimentally by employing the TCD [30], pre-cracked samples [19] and even by considering the atomic scale by means of molecular simulations [15,22]. The fracture tests are realized by pushing the nano-cantilevers towards an indenter which is able to detect the applied load. Table 2 summarizes the load at failure, P_f, as obtained in [30], which should be consulted for additional details on experimental procedures and loading device.

3. Results

The analytical frame developed in Section 2.1 is fully applicable provided the low ratios a/W and ρ/a presented in the previous section (Table 2). By recalling the apparent fracture toughness of a sharp V-notch defined according to [35]

$$K_{Ic}^V = \lambda^\lambda \left[\frac{(2\pi)^{2\lambda-1}}{\beta^2/2} \right]^{1-\lambda} \frac{K_{Ic}^{2(1-\lambda)}}{\sigma_u^{1-2\lambda}}, \tag{11}$$

the dimensionless apparent generalized fracture toughness is plotted in Figure 3 versus the dimensionless notch root radius ρ/l_{ch}, where $l_{ch} = (K_{Ic}/\sigma_u)^2 \approx 5.18$ nm is the so called "Irwin's length" (see Section 2.2 for the mechanical properties). In addition to the geometries considered in the present work, Figure 3 also includes the results for the opening angle $\omega = 0°$ and $\omega = 150°$, as references. When $\rho \to 0$, the fracture toughness ratio correctly tends to 1 since $K_{Ic}^{V,\rho}$ is approaching the value of the sharp ($\rho = 0$) V-notch K_{Ic}^V. Given a fixed value of the normalized notch root radius, the ratio increases as the the opening angle ω decreases, i.e., the lower the notch amplitude, the higher the influence of the radius.

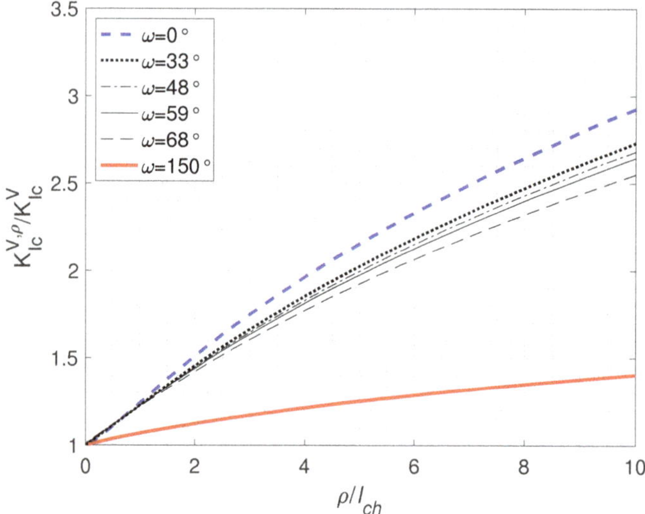

Figure 3. Dimensionless apparent generalized fracture toughness versus dimensionless notch root radius for the opening angles of the nano-cantilever considered in Section 2.2. For the sake of clarity, two additional opening angles are added as references, i.e., $\omega = 0°$ and $\omega = 150°$. Irwin's length $l_{ch} = (K_{Ic}/\sigma_u)^2 \approx 5.18$ nm.

The dimensionless crack extension is plotted in Figure 4. As $\rho \to 0$, the geometry reverts to the sharp V-notch case, with the curves approaching the values provided by the function [35]:

$$l_c^V = \frac{2}{\lambda \beta^2 (2\pi)^{2(1-\lambda)}} l_{ch},\tag{12}$$

which leads to a decreasing crack length for higher notch amplitudes. As ρ increases the element becomes smoother, and each function converges to the asymptotic value $l_c = 2/\pi \left(\frac{K_{Ic}}{1.12\sigma_u}\right)^2$. The most significant deviations from the sharp case are again generally observed as the notch amplitude decreases. Semi-analytical values of crack advancements are also reported in Table 3, showing that for the current geometries all the crack advances were within 10%. Dimensionless values of the notch root radius are reported as well.

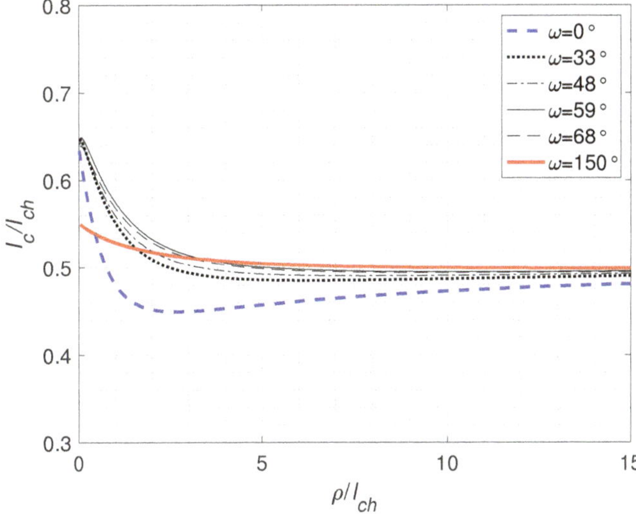

Figure 4. Critical crack advancement versus notch root radius normalized by Irwin's length, $l_{ch} = (K_{Ic}/\sigma_u)^2 \approx 5.18$ nm.

Table 3. Critical crack advancement for the geometries under consideration. The reference crack advancement related to a sharp notch provided by Equation (12) is also reported. Irwin's length is equal to $l_{ch} = (K_{Ic}/\sigma_u)^2 \approx 5.18$ nm.

ω (deg)	ρ/l_{ch}	l_c^V (nm)	l_c (nm)	l_c/l_c^V
33	1.97	3.27	2.64	0.806
48	2.67	3.25	2.62	0.807
59	3.90	3.23	2.60	0.805
68	1.22	3.17	2.83	0.894

Finally, the theoretical FFM estimations based on Figures 3 and 4 are compared with experimental values. In order to obtain the experimental apparent fracture toughness, the stress at the notch root $\sigma_{max} = \sigma_y(0)$ at incipient failure must be evaluated through FEM analysis. Indeed, numerical simulations were already carried out in [30] to study the local stress state and are here exploited, for the sake of simplicity. The experimental values of $K_{Ic}^{V,\rho}$ can be then easily obtained from Equation (13):

$$K_{Ic}^{V,\rho}\text{exp.} = \sigma_{max}\frac{(2\pi r_0)^{1-\lambda}}{[1+\eta_\theta(0)]},\tag{13}$$

and results are summarized in Table 4 and Figure 5. It is clear that FFM accurately estimates the apparent generalized fracture toughness with a negligible error of 4.7% in the worst case ($\omega = 33°$), and an excellent agreement in the best case ($\omega = 48°$).

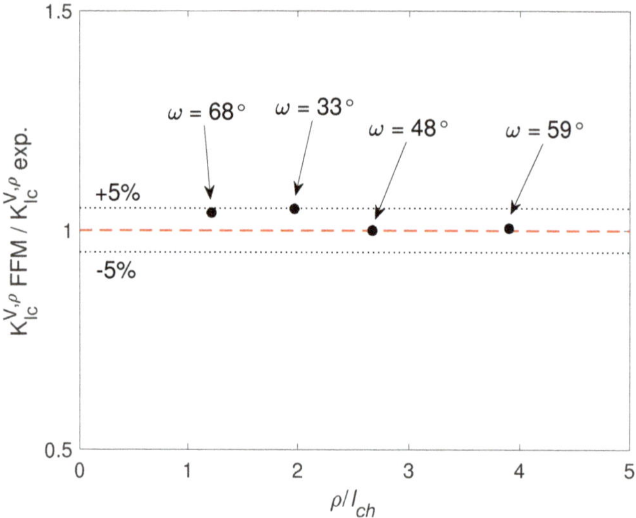

Figure 5. Comparison between experimental (denoted by *exp.*) and Finite Fracture Mechanics (FFM) apparent generalized fracture toughness for the considered nano-cantilevers. The ratio is plotted against the dimensionless notch root radius.

Table 4. Comparison between experimental (denoted by *exp.*) and FFM apparent generalized fracture toughness (MPa m$^{1-\lambda}$) for the considered nano-cantilevers. For the sake of clarity, dimensionless values of the notch root radius are reported as well, where $l_{ch} = (K_{Ic}/\sigma_u)^2 \approx 5.18$ nm.

ω (deg)	ρ/l_{ch}	K_{Ic}^V	$K_{Ic}^{V,\rho}$ FFM	$K_{Ic}^{V,\rho}$ exp.
33	1.97	1.04	1.50	1.43
48	2.67	1.08	1.70	1.70
59	3.90	1.15	2.05	2.04
68	1.22	1.25	1.56	1.50

4. Discussion

The results presented in Section 3 clearly show that the FFM is able to characterize the fracture process of single crystal silicon nanoscale samples. The main advantage of the approach is that, similarly to the TCD [30], it can easily handle notched and cracked geometries. Difficulties in realizing accurate cracks, when dealing with ideal brittle fracture, are indeed well-known and could result in apparent scale dependent fracture toughness. Figure 3 clearly shows that also for a very small notch radius, the apparent generalized fracture toughness deviates from the sharp case. This means that small notch radii cannot be assumed or simplified as sharp. This mistake is commonly made when the FIB processing system (or chemical etching) is used to directly introduced a crack-like notch, based on the assumption that for a small ρ/a ratio the geometry could be treated as a crack. Unfortunately, this assumption does not hold at a very small scale for ideal brittle materials such as the single crystal silicon, that would require, ideally, an atomically sharp crack (ideal Griffith's crack) [22] to determine the proper constant and scale independent fracture toughness. The FFM approach proposed here, instead, provides directly the generalized apparent fracture toughness and gives the possibility to handle different notch radii and opening angles. This conclusion is supported by Figure 5, where the failure assessment is reliable for all the geometries, with negligible differences.

FFM is added therefore to the list of methods that have been proven to be successful when applied at small scale, such as the TCD [30] and the strain energy density [17]. Indeed, these methods share several similarities. Strain energy density and FFM have the same proportionality between the critical notch stress intensity factor and the two fundamental properties i.e., the fracture toughness and the ultimate tensile strength. The difference is that in the FFM the relation depends on notch opening angle only, while in the strain energy density it depends on Poisson's ratio as well [51]. TCD and FFM, instead, both rely on the definition of a critical length scale parameter. While according to the TCD the critical distance is a material constant, the FFM critical crack advancement l_c depends both on the notch radius and amplitude [33]. Figure 4 shows that when the notch radius tends to large values, i.e., smooth samples, all the curves converge towards the same constant value. In this case, the energy requirement provides a constant crack advancement in agreement with the quantized fracture mechanics (QFM) [52], while the stress condition reduces to $\sigma = \sigma_u$ and determines the final failure. When the notch radius tends to zero, the geometry reverts to the sharp V-notch case: In the crack case ($\omega = 0°$), the energy balance reverts to $K_I = K_{Ic}$ and governs the failure mechanism, while the stress requirement leads to the constant critical crack advancement related to the TCD line method [36].

Generally speaking, for all the methods based on a length parameter, the latter can be related to Irwin's length $l_{ch} = (K_{Ic}/\sigma_u)^2$, with $l_{ch} \approx 5.18$ nm in the present work. Recalling the synthesis by means of the TCD presented in [30,36], it could be asserted that all the critical distances are proportional to [53]:

$$\bar{l}_c = \frac{1}{\pi}\left(\frac{K_{Ic}}{\sigma_u}\right)^2 = \frac{1}{\pi}l_{ch}. \tag{14}$$

By substituting numerical values in Equation (14), a critical length $\bar{l}_c \approx 1.65$ nm is obtained. This value obviously agrees with the critical distance evaluated experimentally in [30] i.e., ≈ 1.3–1.9 nm, and it is very close to the critical singular stress field length of 1.5–3 nm at which the continuum-based LEFM breaks down, i.e., 4–5 times the fracture process zone of 0.4–0.6 nm [15,17,22]. Interestingly, all the crack advancements presented in Table 3 are well far from the fracture process zone size. Generalizing, Figure 4 shows that l_c can only vary in a range of ≈ 0.45–0.65 l_{ch}, i.e., 2.3–3.4 nm. Again, these values are larger than the fracture process zone size and the critical length. The results confirm that if that low limit is not reached, continuum-based FFM method is still valid, despite the small scale involved.

At a smaller scale, e.g., atomic scale, extension and generalization of the approach are possible. The FFM, as introduced earlier, is a coupled criterion, i.e., it requires two conditions to be full-filled, one based on stress and the other on energy considerations. The stress criterion could be developed by considering the virial stress, provided that it is accepted as a representation of the mechanical stress at the atomic scale as recently showed in [20], or by employing a traction vector-based stress recently proposed in [54]. The energy criterion, instead, could be defined based on interatomic potential energy as provided in some recent examples based on the energy release rate [22,23] and on the averaged strain energy density [17].

The analysis presented here is limited to mode I loading conditions and to ideal brittle material. In case of semi-brittle materials, experimental conditions could affect the elastic-plastic fracture toughness, and scale dependency of fracture properties may occur [32,55]. This aspect should be carefully considered to define proper mechanical properties and therefore failure conditions of semi-brittle materials.

5. Conclusions

When approaching small scales, the validity of continuum-based LEFM methods is questioned. The present work sheds light on the applicability of the FFM approach by using experimental data available in the literature related to blunt V-notched nano-cantilevers, made of single crystal silicon. The work, therefore, focuses on ideal brittle fracture and mode I loading condition. The main conclusions can be summarised as follows:

- The comparison with experimental tests shows that the FFM is able to predict the failure load of notched nano-cantilevers at the considered small scale (Figure 5). The agreement between experimental and theoretical apparent fracture toughness was excellent for all the opening angles and notch root radii considered, with a maximum discrepancy of 4.7%.
- The validity of the FFM at the considered scale is due to the fact that the low limit of continuum-based LEFM is not yet reached. Indeed, the crack advancements obtained are larger than the fracture process zone of 0.4–0.5 nm evaluated by other authors (Figure 4 and Table 3).
- The critical length $\bar{l}_c \approx 1.65$ nm agrees well with the critical distance and the singular stress field length available in the literature and at which continuum-based LEFM breaks down. The crack advancements obtained here are all well far from this value (Figure 4), further justifying the validity of the method.
- Moreover, for very small notch root radii, the apparent generalized fracture toughness deviates from the sharp case (Figure 3). The deviation is larger for small notch opening angles. When dealing with cracks, if attention is not paid in the realization of an ideally sharp crack-tip, an apparent size dependent fracture toughness may result from experimental tests.
- Based on the evidence presented in the current work, the FFM can be an extremely useful method to predict the static failure of micro and nanodevices made of ideal brittle materials. Further investigations should be done when dealing with semi-brittle materials.
- As further development, the extension of the method to the atomic scale is possible by substituting continuum stress and energy formulations with the virial stress and interatomic potential energy, respectively.

Author Contributions: Conceptualization, P.G.; investigation, P.G.; formal analyses, P.G. and A.S.; writing–original draft preparation, P.G. and A.S.; writing–review and editing, P.G. and A.S.; funding acquisition, P.G. All authors have read and agreed to the published version of the manuscript.

Funding: This research was funded by the Academy of Finland grant number 321244.

Acknowledgments: P.G. is grateful to T. Kitamura, T. Sumigawa (Kyoto University) and Y. Yan (East China University of Science and Technology) for their support during the realization of the experiments.

Conflicts of Interest: The authors declare no conflict of interest.

References

1. Sumigawa, T.; Kitamura, T. In-Situ Mechanical Testing of Nano-Component in TEM. In *Transmission Electron Microscopy*; Maaz, K., Ed.; IntechOpen: Rijeka, Hrvatska, 2012; pp. 355–380.
2. Dehm, G.; Jaya, B.; Raghavan, R.; Kirchlechner, C. Overview on micro- and nanomechanical testing: New insights in interface plasticity and fracture at small length scales. *Acta Mater.* **2018**, *142*, 248–282. [CrossRef]
3. Kheradmand, N.; Rogne, B.R.; Dumoulin, S.; Deng, Y.; Johnsen, R.; Barnoush, A. Small scale testing approach to reveal specific features of slip behavior in BCC metals. *Acta Mater.* **2019**, *174*, 142–152. [CrossRef]
4. Pippan, R.; Wurster, S.; Kiener, D. Fracture mechanics of micro samples: Fundamental considerations. *Mater. Des.* **2018**, *159*, 252–267. [CrossRef]
5. Fang, H.; Shiohara, R.; Sumigawa, T.; Kitamura, T. Size dependence of fatigue damage in sub-micrometer single crystal gold. *Mater. Sci. Eng. A* **2014**, *618*, 416–423. [CrossRef]
6. Shimada, T.; Kitamura, T. Fracture Mechanics at Atomic Scales. In *Advanced Structured Materials*; Altenbach, H., Matsuda, T., Okumura, D., Eds.; Springer International Publishing: Cham, Switzerland, 2015; Volume 64, pp. 379–396. [CrossRef]

7. Kitamura, T.; Sumigawa, T.; Shimada, T.; Lich, L.V. Challenge toward nanometer scale fracture mechanics. *Eng. Fract. Mech.* **2018**, *187*, 33–44. [CrossRef]
8. Bitzek, E.; Kermode, J.R.; Gumbsch, P. Atomistic aspects of fracture. *Int. J. Fract.* **2015**, *191*, 13–30. [CrossRef]
9. Malitckii, E.; Remes, H.; Lehto, P.; Yagodzinskyy, Y.; Bossuyt, S.; Hänninen, H. Strain accumulation during microstructurally small fatigue crack propagation in bcc Fe-Cr ferritic stainless steel. *Acta Mater.* **2018**, *144*, 51–59. [CrossRef]
10. Horstemeyer, M.; Farkas, D.; Kim, S.; Tang, T.; Potirniche, G. Nanostructurally small cracks (NSC): A review on atomistic modeling of fatigue. *Int. J. Fatigue* **2010**, *32*, 1473–1502. [CrossRef]
11. Kupka, D.; Huber, N.; Lilleodden, E.T. A combined experimental-numerical approach for elasto-plastic fracture of individual grain boundaries. *J. Mech. Phys. Solids* **2014**, *64*, 455–467. [CrossRef]
12. Lee, G.H.; Chung, Y.J.; Na, S.M.; Beom, H.G. Atomistic investigation of the T-stress effect on fracture toughness of copper and aluminum single crystals. *J. Mech. Sci. Technol.* **2018**, *32*, 3765–3774. [CrossRef]
13. Sorensen, D.; Hintsala, E.; Stevick, J.; Pischlar, J.; Li, B.; Kiener, D.; Myers, J.C.; Jin, H.; Liu, J.; Stauffer, D.; et al. Intrinsic toughness of the bulk-metallic glass Vitreloy 105 measured using micro-cantilever beams. *Acta Mater.* **2020**, *183*, 242–248. [CrossRef]
14. Yan, Y.; Sumigawa, T.; Wang, X.; Chen, W.; Xuan, F.; Kitamura, T. Fatigue curve of microscale single-crystal copper: An in situ SEM tension-compression study. *Int. J. Mech. Sci.* **2020**, *171*, 105361. [CrossRef]
15. Shimada, T.; Ouchi, K.; Chihara, Y.; Kitamura, T. Breakdown of Continuum Fracture Mechanics at the Nanoscale. *Sci. Rep.* **2015**, *5*, 8596. [CrossRef] [PubMed]
16. Gallo, P.; Sumigawa, T.; Shimada, T.; Yan, Y.; Kitamura, T. Investigation into the Breakdown of Continuum Fracture Mechanics at the Nanoscale: Synthesis of Recent Results on Silicon. In *Proc. First Int. Conf. Theor. Appl. Exp. Mech.*; Gdoutos, E.E., Ed.; Springer International Publishing: Cham, Switzerland, 2019; pp. 205–210. [CrossRef]
17. Gallo, P.; Hagiwara, Y.; Shimada, T.; Kitamura, T. Strain energy density approach for brittle fracture from nano to macroscale and breakdown of continuum theory. *Theor. Appl. Fract. Mech.* **2019**, *103*, 102300. [CrossRef]
18. Liu, F.; Tang, Q.; Wang, T.C. Intrinsic Notch Effect Leads to Breakdown of Griffith Criterion in Graphene. *Small* **2017**, *1700028*, 1–8. [CrossRef]
19. Sumigawa, T.; Ashida, S.; Tanaka, S.; Sanada, K.; Kitamura, T. Fracture toughness of silicon in nanometer-scale singular stress field. *Eng. Fract. Mech.* **2015**, *150*, 161–167. [CrossRef]
20. Gallo, P. On the Crack-Tip Region Stress Field in Molecular Systems: The Case of Ideal Brittle Fracture. *Adv. Theory Simulations* **2019**, *2*, 1900146. [CrossRef]
21. Singh, G.; Kermode, J.R.; De Vita, A.; Zimmerman, R.W. Validity of linear elasticity in the crack-tip region of ideal brittle solids. *Int. J. Fract.* **2014**, *189*, 103–110. [CrossRef]
22. Sumigawa, T.; Shimada, T.; Tanaka, S.; Unno, H.; Ozaki, N.; Ashida, S.; Kitamura, T. Griffith Criterion for Nanoscale Stress Singularity in Brittle Silicon. *ACS Nano* **2017**, *11*, 6271–6276. [CrossRef]
23. Huang, K.; Shimada, T.; Ozaki, N.; Hagiwara, Y.; Sumigawa, T.; Guo, L.; Kitamura, T. A unified and universal Griffith-based criterion for brittle fracture. *Int. J. Solids Struct.* **2017**, *128*, 67–72. [CrossRef]
24. Yin, H.; Qi, H.J.; Fan, F.; Zhu, T.; Wang, B.; Wei, Y. Griffith criterion for brittle fracture in graphene. *Nano Lett.* **2015**, *15*, 1918–1924. [CrossRef] [PubMed]
25. Huang, K.; Sumigawa, T.; Kitamura, T. Load-dependency of damage process in tension-compression fatigue of microscale single-crystal copper. *Int. J. Fatigue* **2020**, *133*, 105415. [CrossRef]
26. Gallo, P.; Sumigawa, T.; Kitamura, T.; Berto, F. Static assessment of nanoscale notched silicon beams using the averaged strain energy density method. *Theor. Appl. Fract. Mech.* **2018**, *95*, 261–269. [CrossRef]
27. Gallo, P.; Sumigawa, T.; Kitamura, T. Experimental characterization at nanoscale of single crystal silicon fracture toughness. *Frat. Ed Integrità Strutt.* **2019**, *13*, 408–415. [CrossRef]
28. Taylor, D. The Theory of Critical Distances: A link to micromechanisms. *Theor. Appl. Fract. Mech.* **2017**, *90*, 228–233. [CrossRef]
29. Susmel, L.; Taylor, D. The theory of critical distances to predict static strength of notched brittle components subjected to mixed-mode loading. *Eng. Fract. Mech.* **2008**, *75*, 534–550. [CrossRef]
30. Gallo, P.; Yan, Y.; Sumigawa, T.; Kitamura, T. Fracture Behavior of Nanoscale Notched Silicon Beams Investigated by the Theory of Critical Distances. *Adv. Theory Simulat.* **2018**, *1*, 1700006. [CrossRef]

31. Ando, T.; Li, X.; Nakao, S.; Kasai, T.; Tanaka, H.; Shikida, M.; Sato, K. Fracture toughness measurement of thin-film silicon. *Fatigue Fract. Eng. Mater. Struct.* **2005**, *28*, 687–694. [CrossRef]

32. Saxena, A.K.; Brinckmann, S.; Völker, B.; Dehm, G.; Kirchlechner, C. Experimental conditions affecting the measured fracture toughness at the microscale: Notch geometry and crack extension measurement. *Mater. Des.* **2020**, 108582. [CrossRef]

33. Liu, Y.; Deng, C.; Gong, B. Discussion on equivalence of the theory of critical distances and the coupled stress and energy criterion for fatigue limit prediction of notched specimens. *Int. J. Fatigue* **2020**, *131*, 105326. [CrossRef]

34. Leguillon, D. Strength or toughness? A criterion for crack onset at a notch. *Eur. J. Mech. A/Solids* **2002**, *21*, 61–72. [CrossRef]

35. Carpinteri, A.; Cornetti, P.; Pugno, N.; Sapora, A.; Taylor, D. A finite fracture mechanics approach to structures with sharp V-notches. *Eng. Fract. Mech.* **2008**, *75*, 1736–1752. [CrossRef]

36. Carpinteri, A.; Cornetti, P.; Sapora, A. Brittle failures at rounded V-notches: A finite fracture mechanics approach. *Int. J. Fract.* **2011**, *172*, 1–8. [CrossRef]

37. Sapora, A.; Cornetti, P.; Carpinteri, A.; Firrao, D. An improved Finite Fracture Mechanics approach to blunt V-notch brittle fracture mechanics: Experimental verification on ceramic, metallic, and plastic materials. *Theor. Appl. Fract. Mech.* **2015**, *78*, 20–24. [CrossRef]

38. Mantič, V. Interface crack onset at a circular cylindrical inclusion under a remote transverse tension. Application of a coupled stress and energy criterion. *Int. J. Solids Struct.* **2009**, *46*, 1287–1304. [CrossRef]

39. Felger, J.; Rosendahl, P.; Leguillon, D.; Becker, W. Predicting crack patterns at bi-material junctions: A coupled stress and energy approach. *Int. J. Solids Struct.* **2019**, *164*, 191–201. [CrossRef]

40. Yosibash, Z.; Mittelman, B. A 3-D failure initiation criterion from a sharp V-notch edge in elastic brittle structures. *Eur. J. Mech. A/Solids* **2016**, *60*, 70–94. [CrossRef]

41. Brochard, L.; Tejada, I.G.; Sab, K. From yield to fracture, failure initiation captured by molecular simulation. *J. Mech. Phys. Solids* **2016**, *95*, 632–646. [CrossRef]

42. Sapora, A.; Cornetti, P. Crack onset and propagation stability from a circular hole under biaxial loading. *Int. J. Fract.* **2018**, *214*, 97–104. [CrossRef]

43. Sapora, A.; Cornetti, P.; Campagnolo, A.; Meneghetti, G. Fatigue limit: Crack and notch sensitivity by Finite Fracture Mechanics. *Theor. Appl. Fract. Mech.* **2020**, *105*, 102407. [CrossRef]

44. Torabi, A.R.; Berto, F.; Sapora, A. Finite Fracture Mechanics Assessment in Moderate and Large Scale Yielding Regimes. *Metals (Basel)* **2019**, *9*, 602. [CrossRef]

45. Cornetti, P.; Sapora, A.; Carpinteri, A. Short cracks and V-notches: Finite Fracture Mechanics vs. Cohesive Crack Model. *Eng. Fract. Mech.* **2016**, *168*, 2–12. [CrossRef]

46. Cornetti, P.; Muñoz-Reja, M.; Sapora, A.; Carpinteri, A. Finite fracture mechanics and cohesive crack model: Weight functions vs. cohesive laws. *Int. J. Solids Struct.* **2019**, *156–157*, 126–136. [CrossRef]

47. Doitrand, A.; Estevez, R.; Leguillon, D. Comparison between cohesive zone and coupled criterion modeling of crack initiation in rhombus hole specimens under quasi-static compression. *Theor. Appl. Fract. Mech.* **2019**, *99*, 51–59. [CrossRef]

48. Gentieu, T.; Jumel, J.; Catapano, A.; Broughton, J. Size effect in particle debonding: Comparisons between finite fracture mechanics and cohesive zone model. *J. Compos. Mater.* **2019**, *53*, 1941–1954. [CrossRef]

49. Filippi, S.; Lazzarin, P.; Tovo, R. Developments of some explicit formulas useful to describe elastic stress fields ahead of notches in plates. *Int. J. Solids Struct.* **2002**, *39*, 4543–4565. [CrossRef]

50. Sapora, A.; Cornetti, P.; Carpinteri, A. Cracks at rounded V-notch tips: An analytical expression for the stress intensity factor. *Int. J. Fract.* **2014**, *187*, 285–291. [CrossRef]

51. Lazzarin, P.; Campagnolo, A.; Berto, F. A comparison among some recent energy- and stress-based criteria for the fracture assessment of sharp V-notched components under Mode I loading. *Theor. Appl. Fract. Mech.* **2014**, *71*, 21–30. [CrossRef]

52. Pugno, N.M.; Ruoff, R.S. Quantized fracture mechanics. *Philos. Mag.* **2004**, *84*, 2829–2845. [CrossRef]

53. Cicero, S.; Fuentes, J.; Procopio, I.; Madrazo, V.; González, P. Critical Distance Default Values for Structural Steels and a Simple Formulation to Estimate the Apparent Fracture Toughness in U-Notched Conditions. *Metals (Basel)* **2018**, *8*, 871. [CrossRef]

54. Yang, J.; Komvopoulos, K. A stress analysis method for molecular dynamics systems. *Int. J. Solids Struct.*
     2020. [CrossRef]
55. Wang, Y.; Fritz, R.; Kiener, D.; Zhang, J.; Liu, G.; Kolednik, O.; Pippan, R.; Sun, J. Fracture behavior and
     deformation mechanisms in nanolaminated crystalline/amorphous micro-cantilevers. *Acta Mater.* **2019**,
     180, 73–83. [CrossRef]

© 2020 by the authors. Licensee MDPI, Basel, Switzerland. This article is an open access
article distributed under the terms and conditions of the Creative Commons Attribution
(CC BY) license (http://creativecommons.org/licenses/by/4.0/).

Article

Peridynamic Modeling of Mode-I Delamination Growth in Double Cantilever Composite Beam Test: A Two-Dimensional Modeling Using Revised Energy-Based Failure Criteria

Xiao-Wei Jiang [1,2], Shijun Guo [2], Hao Li [2] and Hai Wang [1,*]

[1] School of Aeronautics and Astronautics, Shanghai Jiao Tong University, Shanghai 200240, China; jiangxiaowei@sjtu.edu.cn
[2] Centre of Aeronautics, Cranfield University, Cranfield, MK43 0AL, UK; s.guo@cranfield.ac.uk (S.G.); hao.li@cranfield.ac.uk (H.L.)
*   Correspondence: wanghai601@sjtu.edu.cn; Tel.: +86-021-34206395

Received: 24 December 2018; Accepted: 12 February 2019; Published: 15 February 2019

Abstract: This study presents a two-dimensional ordinary state-based peridynamic (OSB PD) modeling of mode-I delamination growth in a double cantilever composite beam (DCB) test using revised energy-based failure criteria. The two-dimensional OSB PD composite model for DCB modeling is obtained by reformulating the previous OSB PD lamina model in x–z direction. The revised energy-based failure criteria are derived following the approach of establishing the relationship between critical bond breakage work and energy release rate. Loading increment convergence analysis and grid spacing influence study are conducted to investigate the reliability of the present modeling. The peridynamic (PD) modeling load–displacement curve and delamination growth process are then quantitatively compared with experimental results obtained from standard tests of composite DCB samples, which show good agreement between the modeling results and experimental results. The PD modeling delamination growth process damage contours are also illustrated. Finally, the influence of the revised energy-based failure criteria is investigated. The results show that the revised energy-based failure criteria improve the accuracy of the PD delamination modeling of DCB test significantly.

Keywords: peridynamics; composite; ordinary state-based; double cantilever composite beam (DCB); delamination

---

1. Introduction

Carbon fiber-reinforced polymer (CFRP) composite materials have been widely used in aerospace structures due to their high specific stiffness/strength, low coefficient of thermal expansion, and excellent fatigue resistance. In the design procedure of composite structures, testing and analysis are both needed due to the overall consideration of cost and reliability. Laminate delamination is one of the main failure modes of composite structures and, hence, has become a major concern of CFRP composite design and analysis for many years [1–5]. Currently, the most widely used analysis method for capturing delamination initiation and growth of CFRP composites is based on the framework of finite element method (FEM), such as cohesive zone element (CZE) and virtual crack closure technique (VCCT). Although these techniques have been successfully applied to many delamination problems of CFRP composites [6–8], they usually have to preset the delamination growth path, which is difficult for many real engineering problems. Besides, as stated in [9–12], these conventional analysis methods based on classical continuum mechanics require that the displacement field of body be continuously differential in space. This requirement is in contradiction with inherent spatial discontinuity existing in delamination growth problems.

As an alternative to conventional analysis method, Silling et al. [13–15] introduced the peridynamic (PD) theory of solid mechanics, which attempts to unite the mathematical modeling of continuous media, cracks, and particles within a single framework. Peridynamic theory replaces the partial differential equation of the classical theory of solid mechanics with integral or integral-differential equations, and "spontaneous" formation of fracture and damage of composites could be simulated without any prior knowledge of damage path [16–19]. Peridynamic theory has been successfully applied to capture the delamination damage of CFRP composites. The interlayer delamination damage patterns of composite laminate under low velocity impact were given by Askari et al. [11] and Xu et al. [10]. The delamination damages between each layer of composite laminates with notch or open hole were presented by Hu et al. [20–23]. "Spontaneous" delamination damages were captured by peridynamic theory in these works, without any presetting of delamination path or refinement of the delamination front mesh.

Although the above works have proved the potential advantages of peridynamic theory in delamination modeling of CFRP composites, currently, few studies have focused on the quantitative delamination growth modeling of CFRP composite by peridynamics. Double cantilever composite beam (DCB) test provides a good benchmark example for validating an approach's basic ability in modeling mode-I delamination in composites. It is meaningful to conduct DCB modeling work to examine the reliability of an approach in modeling structural delamination problem, such as for CZE [24,25] and VCCT [26]. Hu et al. [12] developed an energy-based approach to simulate delamination in composites using bond-based peridynamics. Delamination growth in double cantilever beam (DCB) test is simulated and the convex delamination front was well captured.

In the present study, a two-dimensional ordinary state-based peridynamic (OSB PD) modeling of mode-I delamination growth of CFRP composite DCB test is presented. Double cantilever beam (DCB) test is the basic experiment to measure mode-I delamination fracture toughness, G_{IC}, of CFRP composites. This simple loading case provides a good benchmark example to quantitatively verify the ability of peridynamic theory in modeling delamination growth of CFRP composites. The two-dimensional OSB PD composite model used in the present study for DCB modeling is obtained by reformulating the previous OSB PD lamina model [27] in the x–z direction. The revised energy-based failure criteria are derived following the approach proposed by Silling et al. [15] of establishing the relationship between critical bond breakage work and energy release rate. Loading increment convergence analysis and grid spacing influence study are conducted. Comparisons of load–displacement curve and delamination growth process between PD modeling results and experimental results are shown. Delamination growth process damage contours are illustrated. Numerical results using revised and original energy-based failure criteria are also compared.

2. Two-Dimensional Ordinary State-Based Peridynamic Model for DCB Modeling

2.1. Governing Equation

The two-dimensional OSB PD composite model used in the present study for DCB modeling is obtained by reformulating the previous OSB PD lamina model (Madenci and Oterkus [27]) in x–z direction. The detailed derivation process is given in Appendix A. The governing equation of the present two-dimensional OSB PD composite model can be written as

$$\rho_{(k)}\ddot{\mathbf{u}}_{(k)} = \sum_{j=1}^{\infty} [t_{(k)(j)} - \mathbf{t}_{(j)(k)}]V_{(j)} + \mathbf{b}_{(k)}, \tag{1}$$

where $\rho_{(k)}$ is the density of material point $\mathbf{x}_{(k)}$, $\ddot{\mathbf{u}}_{(k)}$ is instantaneous acceleration of $\mathbf{x}_{(k)}$, as shown in Figure 1, $\mathbf{b}_{(k)}$ is the external load density, and $\mathbf{t}_{(k)(j)}$ and $\mathbf{t}_{(j)(k)}$ are PD force density between $\mathbf{x}_{(k)}$ and $\mathbf{x}_{(j)}$. The PD force density can be expressed as

$$\mathbf{t}_{(k)(j)} = A_{(k)(j)} \frac{\mathbf{y}_{(j)} - \mathbf{y}_{(k)}}{\left|\mathbf{y}_{(j)} - \mathbf{y}_{(k)}\right|}, \tag{2}$$

$$\mathbf{t}_{(j)(k)} = B_{(j)(k)} \frac{\mathbf{y}_{(k)} - \mathbf{y}_{(j)}}{\left|\mathbf{y}_{(k)} - \mathbf{y}_{(j)}\right|}, \tag{3}$$

with

$$A_{(k)(j)} = 2ad \frac{\delta}{\left|\mathbf{x}_{(j)} - \mathbf{x}_{(k)}\right|} \Lambda_{(k)(j)} \theta_{(k)} + 2\delta b s_{(k)(j)} + 2\delta(\mu_F b_F + \mu_Z b_Z) s_{(k)(j)}, \tag{4}$$

$$B_{(j)(k)} = 2ad \frac{\delta}{\left|\mathbf{x}_{(k)} - \mathbf{x}_{(j)}\right|} \Lambda_{(j)(k)} \theta_{(j)} + 2\delta b s_{(j)(k)} + 2\delta(\mu_F b_F + \mu_Z b_Z) s_{(j)(k)}, \tag{5}$$

and

$$s_{(k)(j)} = \frac{\left|\mathbf{y}_{(j)} - \mathbf{y}_{(k)}\right| - \left|\mathbf{x}_{(j)} - \mathbf{x}_{(k)}\right|}{\left|\mathbf{x}_{(j)} - \mathbf{x}_{(k)}\right|}, \tag{6}$$

and

$$\mu_F = \begin{cases} 1 & (\mathbf{x}_{(j)} - \mathbf{x}_{(k)}) \parallel \text{fiber direction} \\ 0 & \text{otherwise} \end{cases}, \tag{7}$$

$$\mu_Z = \begin{cases} 1 & (\mathbf{x}_{(j)} - \mathbf{x}_{(k)}) \perp \text{fiber direction} \\ 0 & \text{otherwise} \end{cases}, \tag{8}$$

where $s_{(k)(j)}$ is the stretch of bonds, and δ is the radius of the horizon zone. The direction cosines of the relative position vectors between the material points $\mathbf{x}_{(k)}$ and $\mathbf{x}_{(j)}$ in the undeformed and deformed states are defined as

$$\Lambda_{(k)(j)} = \frac{\mathbf{y}_{(j)} - \mathbf{y}_{(k)}}{\left|\mathbf{y}_{(j)} - \mathbf{y}_{(k)}\right|} \cdot \frac{\mathbf{x}_{(j)} - \mathbf{x}_{(k)}}{\left|\mathbf{x}_{(j)} - \mathbf{x}_{(k)}\right|}. \tag{9}$$

The PD dilatation $\theta_{(k)}$ can be expressed as

$$\theta_{(k)} = d \sum_{j=1}^{\infty} \delta s_{(k)(j)} \Lambda_{(k)(j)} V_{(j)}. \tag{10}$$

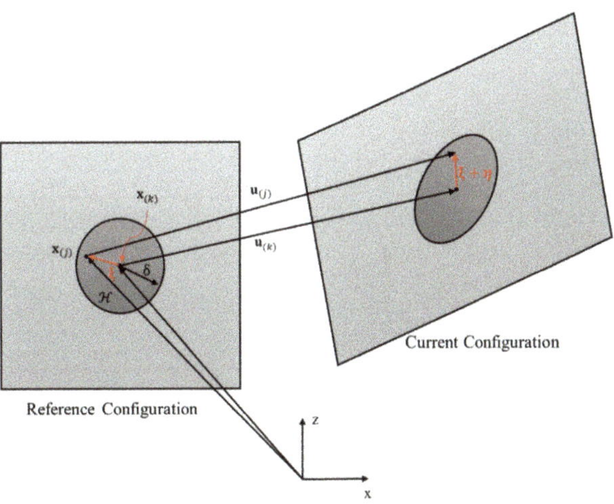

Figure 1. Peridynamic model notations.

The PD material parameters a, d characterize the effect of dilatation, and b, b_F, b_Z are associated with deformation of material points in arbitrary directions, fiber direction, and thickness direction, respectively. These parameters are related to material properties of CFRP composite and horizon radius. The detailed derivation procedures to get these PD material parameters are illustrated in Appendix A.2. These PD material parameters are related to composite material parameters as

$$a = \frac{1}{2}(Q_{13} - Q_{55}), \tag{11}$$

$$d = \frac{2}{\pi\delta^3}, \tag{12}$$

$$b = \frac{6Q_{55}}{\pi\delta^4}, \tag{13}$$

$$b_F = \frac{Q_{11} - Q_{13} - 2Q_{55}}{2\delta \sum\limits_{j=1}^{J} \left| x_{(j)} - x_{(k)} \right| V_{(j)}}, \tag{14}$$

$$b_Z = \frac{Q_{33} - Q_{13} - 2Q_{55}}{2\delta \sum\limits_{j=1}^{J} \left| x_{(j)} - x_{(k)} \right| V_{(j)}}, \tag{15}$$

where Q_{11}, Q_{33}, Q_{13}, and Q_{55} are coefficients of stiffness matrix $[Q]$ presented in Appendix A.1.

2.2. Revised Energy-Based Failure Criteria

Following the approach for deriving the relationship between critical bond breakage work and energy release rate by Silling et al. [15], revised two-dimensional energy-based failure criteria for mode-I delamination growth of CFRP composites are proposed. The influence of the revised energy-based failure criteria will be investigated in Section 4.3.

This approach assumes that the energy consumed by a growing delamination front equals the work required, per unit delamination front area, to separate two halves of a body across a plane (Figure 2). Suppose a plane A separates two halves of a two-dimensional body B into B_+ and B_-. The delamination front area a is on the plane. Consider a mode-I delamination motion with velocity field on Figure 2. The total energy E absorbed by P in this motion is

$$E = \int_0^t W_{abs}(P)dt' = \int_0^{t_0} \int_P \int_B t_z(v_z' - v_z)dV'dVdt. \tag{16}$$

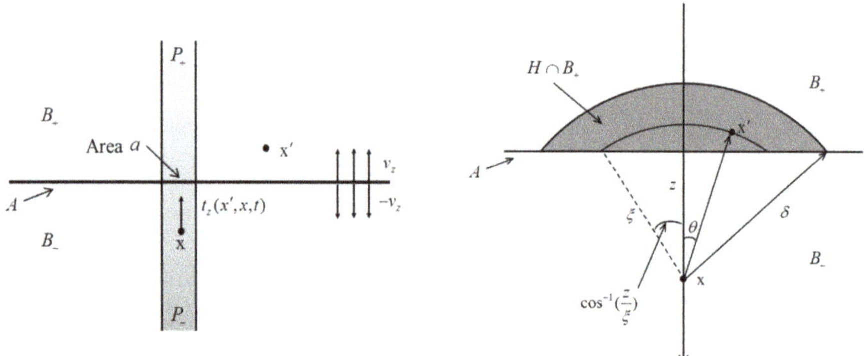

Figure 2. Computation of total energy absorbed by P for delamination front area a.

The assumed critical bond breakage work w_{IC} in this motion is

$$\int_0^{t_0} t_z(v_z{}' - v_z)dt = w_{\text{IC}}. \tag{17}$$

Therefore,

$$
\begin{aligned}
E \;&= w_{\text{IC}} \int_P \int_B dV'dV = w_{\text{IC}}\left(\int_{P_-}\int_{B_+\backslash P_+} dV'dV + \int_{P_+}\int_{B_-\backslash P_-} dV'dV + \int_{P_-}\int_{P_+} dV'dV\right) \\
&= w_{\text{IC}}\left(2\int_{P_-}\int_{B_+} dV'dV - \int_{P_-}\int_{P_+} dV'dV\right) \\
&= w_{\text{IC}}\left(2a\int_0^\delta \int_z^\delta \int_{-\cos^{-1}(z/\xi)}^{\cos^{-1}(z/\xi)} \xi d\theta d\xi dz - a\int_0^\delta a(\delta - z)dz\right) \\
&= w_{\text{IC}}\left(\tfrac{4}{3}a\delta^3 - \tfrac{1}{2}a^2\delta^2\right)
\end{aligned}
\tag{18}
$$

For Equation (18), as stated by Silling et al. [15], bonds that do not cross plane A have no contributions to the integrand. However, in the present study, bonds that cross plane A are separated into three parts: bonds connecting P_- and $B_+\backslash P_+$, bonds connecting P_+ and $B_-\backslash P_-$, and bonds connecting P_- and P_+, as shown in Figure 2. The bond breakage work between P_- and P_+ is calculated only once compared with the original procedure in [15]. Adding these three parts of the bond breakage work together gives the revised total absorbed energy E shown in Equation (18).

Assuming this total absorbed energy equals the critical energy release rate times the area of delamination front,

$$G_{\text{IC}} = E/a = w_{\text{IC}}\left(\frac{4}{3}\delta^3 - \frac{1}{2}a\delta^2\right). \tag{19}$$

Thus, the critical bond breakage work for mode-I delamination is related to critical energy release rate,

$$w_{\text{IC}} = \frac{G_{\text{IC}}}{\frac{4}{3}\delta^3 - \frac{1}{2}a\delta^2}. \tag{20}$$

For two-dimensional numerical modeling, the delamination front area a can be set as the discretized grid spacing dx,

$$w_{\text{IC}} = \frac{G_{\text{IC}}}{\frac{4}{3}\delta^3 - \frac{1}{2}dx\delta^2}. \tag{21}$$

From the above derivation, revised energy-based failure criteria for mode-I delamination growth are proposed as

$$\left|\frac{w_{\text{I}}}{w_{\text{IC}}}\right| \geq 1, \; w_{\text{IC}} = \frac{G_{\text{IC}}}{\frac{4}{3}\delta^3 - \frac{1}{2}dx\delta^2}, \; w_{\text{I}} = \int_0^t t_z(v_z{}' - v_z)dt = t_z(w' - w). \tag{22}$$

Local damage at a material point is defined as the weighted ratio of the number of eliminated interactions to the total number of initial interactions of the material point with its family members [28]. The status variable for delamination damage, μ, is defined as

$$\mu = \begin{cases} 1, & \left|\frac{w_{\text{I}}}{w_{\text{IC}}}\right| < 1 \quad \text{no damage} \\[2mm] 0, & \left|\frac{w_{\text{I}}}{w_{\text{IC}}}\right| \geq 1 \quad \text{damage} \end{cases}. \tag{23}$$

The delamination damage between ply (n) and (n + 1) at a point can be quantified as

$$
\begin{cases}
\varphi^{(n)}_{\text{out-of-plane\_upper}} = 1 - \dfrac{\displaystyle\sum_{j=1}^{N^{(upper)}_{(k)}} \mu^{(n)(m)}_{(k)(j)}}{N^{(upper)}_{(k)}} \\[2em]
\varphi^{(n+1)}_{\text{out-of-plane\_lower}} = 1 - \dfrac{\displaystyle\sum_{j=1}^{N^{(lower)}_{(k)}} \mu^{(n)(m)}_{(k)(j)}}{N^{(lower)}_{(k)}} \\[2em]
\varphi^{(n)(n+1)}_{\text{delamination}} = \frac{1}{2}\left(\varphi^{(n)}_{\text{out-of-plane\_upper}} + \varphi^{(n+1)}_{\text{out-of-plane\_lower}}\right).
\end{cases}
\tag{24}
$$

2.3. Numerical Implementation

Although the peridynamic governing equation is in dynamic form, it can still be used to solve quasi-static or static problems [29–31]. Adaptive dynamic relaxation (ADR) method [29] is currently the most widely used method to solve quasi-static or static problems for PD. ADR method is particularly effective for solving highly nonlinear problems, including geometric and material nonlinearities [31].

According to the ADR method, Equation (1) at the nth iteration can be rewritten as

$$
\ddot{\mathbf{U}}^{n}(\mathbf{X},t^{n}) + c^{n}\dot{\mathbf{U}}^{n}(\mathbf{X},t^{n}) = \mathbf{D}^{-1}\mathbf{F}^{n}(\mathbf{U}^{n},\mathbf{U}'^{n},\mathbf{X},\mathbf{X}'),
\tag{25}
$$

where $\mathbf{D}$ is the fictitious diagonal density matrix and c is the damping coefficient which can be expressed by

$$
c^{n} = 2\sqrt{((\mathbf{U}^{n})^{T}\, {}^{1}\mathbf{K}^{n}\mathbf{U}^{n})/((\mathbf{U}^{n})^{T}\mathbf{U}^{n})},
\tag{26}
$$

in which ${}^{1}\mathbf{K}^{n}$ is the diagonal "local" stiffness matrix, which is given as

$$
{}^{1}K^{n}_{ii} = -(F^{n}_{i}/\lambda_{ii} - F^{n-1}_{i}/\lambda_{ii})/(\Delta t \dot{u}^{n-1/2}_{i}),
\tag{27}
$$

where F^{n}_{i} is the value of force vector $\mathbf{F}^{n}$ at material point $\mathbf{x}$, which includes both the peridynamic force state vector and external forces, and λ_{ii} are the diagonal elements of $\mathbf{D}$ which should be large enough to avoid numerical divergence.

By utilizing central-difference explicit integration, displacements and velocities for the next time step can be obtained:

$$
\dot{\mathbf{U}}^{n+1/2} = \frac{((2 - c^{n}\Delta t)\dot{\mathbf{U}}^{n-1/2} + 2\Delta t \mathbf{D}^{-1}\mathbf{F}^{n})}{(2 + c^{n}\Delta t)}
\tag{28}
$$

and

$$
\dot{\mathbf{U}}^{n+1} = \mathbf{U}^{n} + \Delta t \dot{\mathbf{U}}^{n+1/2}.
\tag{29}
$$

To start the iteration process, we assume that $\mathbf{U}^{0} \neq 0$ and $\dot{\mathbf{U}}^{0} = 0$, so the integration can be started by

$$
\dot{\mathbf{U}}^{1/2} = \frac{\Delta t \mathbf{D}^{-1}\mathbf{F}^{0}}{2}.
\tag{30}
$$

Due to the large computational amount of PD model, GPU-parallel computing is introduced. The PGI CUDA FORTRAN compiler, PGI/17.10 Community Edition, is used for compiling. The GPU node at Cranfield University Delta HPC Cluster is applied for running the GPU-parallel program. The GPU block threads are fixed to 256, and the number of blocks depend on the total number of parallel processes [32].

3. Experimental Setup and Results

In order to validate the reliability of the present two-dimensional OSB PD composite model, a CFRP double cantilever beam (DCB) test is conducted. The CFRP composite DCB test setup is shown in Figure 3a. The experiment was conducted in accordance with ASTM standard D5528-13 [33]. The DCB test specimens are bonded with two piano hinges at the initial delamination end. White paint marking is sprayed on the side of the specimen for visualizing the delamination front position. As shown in Figure 4, the length of the specimen is 167 mm, with 25 mm width and 4 mm thickness. The layup of the specimen is $[0]_{32}$, with 0.125 mm/ply. Initial delamination $a_0 = 50$ mm is preset on the mid-plane of the specimen, between Layer 16 and 17. In this experiment, five DCB specimens are tested. The loading velocity is 1 mm/min. The resulting load–displacement curves of the specimens are presented in Figure 12. The delamination growth of each specimen is also recorded in Table 1. Modified beam theory (MBT) method from ASTM standard D5528-13 was used to analyze the experimental data, and the resulting average G_{IC} of the CFRP composite is shown in Table 2.

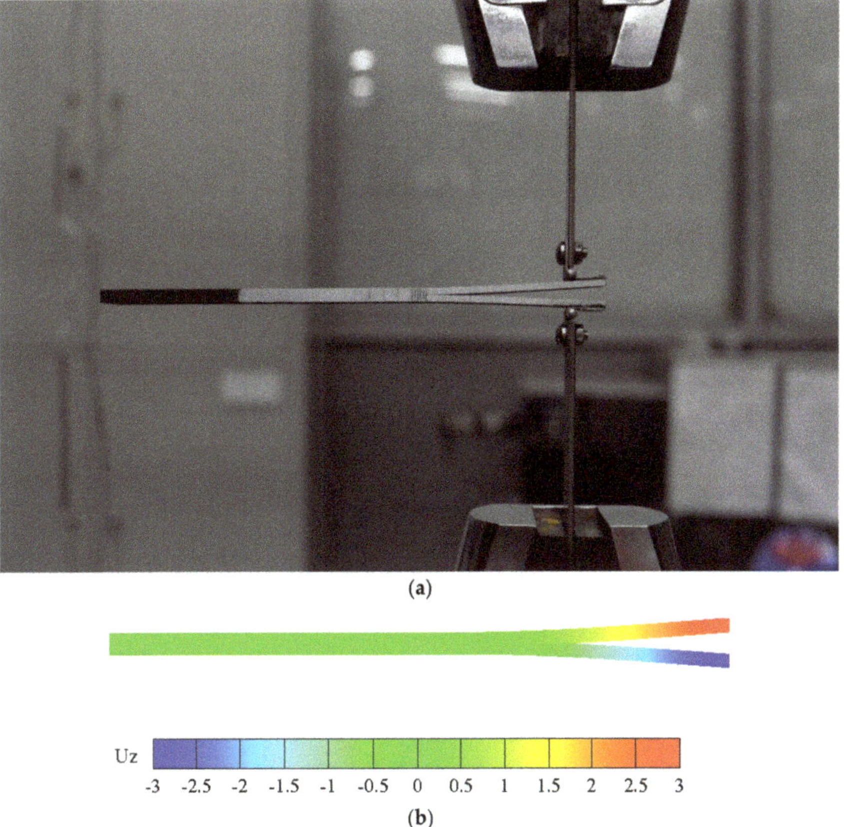

Figure 3. Carbon fiber-reinforced polymer (CFRP) double cantilever beam (DCB) test and two-dimensional PD modeling. (**a**) Double cantilever composite beam (DCB) test setup, (**b**) peridynamic (PD) modeling deformed displacement field in thickness direction (mm).

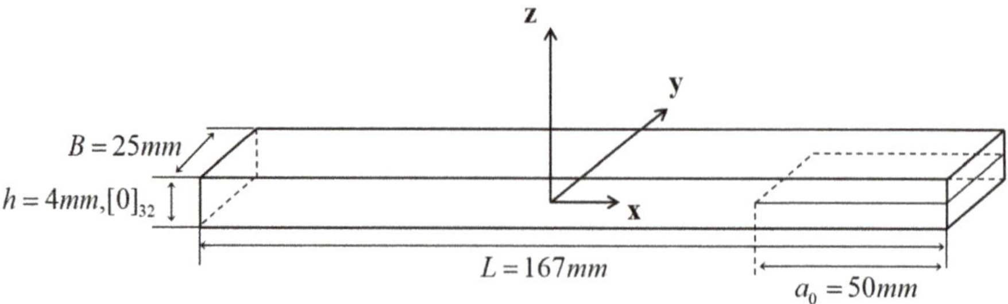

Figure 4. CFRP double cantilever beam (DCB) test specimen.

Table 1. Experimental delamination growth process observed in DCB test (DELL: delamination length).

DELL (mm)	DCB-1 Disp. (mm)	DCB-1 Load (N)	DCB-2 Disp. (mm)	DCB-2 Load (N)	DCB-3 Disp. (mm)	DCB-3 Load (N)	DCB-4 Disp. (mm)	DCB-4 Load (N)	DCB-5 Disp. (mm)	DCB-5 Load (N)
50	4.77	88.04	5.07	97.22	5.52	86.16	5.27	94.29	4.79	91.07
51	5.17	87.59	5.23	86.22	5.59	79.09	5.27	94.29	5.24	91.09
52	5.18	85.84	5.26	82.09	5.60	77.87	5.75	89.35	5.34	90.19
53	5.20	85.23	5.29	81.60	5.62	77.53	5.75	89.35	5.62	92.44
54	5.26	85.60	5.30	80.78	5.65	72.54	5.75	89.35	5.62	92.44
55	5.54	84.56	5.34	80.10	5.67	69.59	5.96	81.77	6.02	93.91
60	6.15	62.19	5.52	77.25	5.89	55.74	6.47	74.20	6.42	66.33
65	6.79	68.00	6.15	66.52	6.03	47.80	7.13	62.66	8.12	73.22

Table 2. Material properties of CFRP composite DCB specimen.

E_1 (GPa)	E_2 (GPa)	G_{12} (GPa)	v_{12}	G_{IC} (N/mm)
127	7.9	4.2	0.32	0.512

4. Numerical Results and Discussion

The double cantilever composite beam test described in Section 3 is modeled using the two-dimensional OSB PD composite model in Section 2. The DCB specimen shown in Figure 4 is made of CFRP composite with material properties shown in Table 2. The horizon of the present two-dimensional OSB PD model is $\delta = 3$ dx, dx is the grid spacing. The boundary conditions for PD DCB model are shown in Figure 5. Displacement boundary conditions are applied at the loading end. The loading increment for each iteration step is noted as dL. No-fail zones are set for preventing possible premature failure due to boundary effects [12]. The modeling flowchart is shown in Figure 6. For each loading step, firstly, a displacement boundary condition dL is applied; secondly, ADR is used to get the static results; thirdly, the revised energy-based failure criteria derived in Section 2.2 are used to check bond failure; and finally, a new loading step is applied. The two-dimensional PD modeling deformed displacement field in thickness direction is shown in Figure 3b.

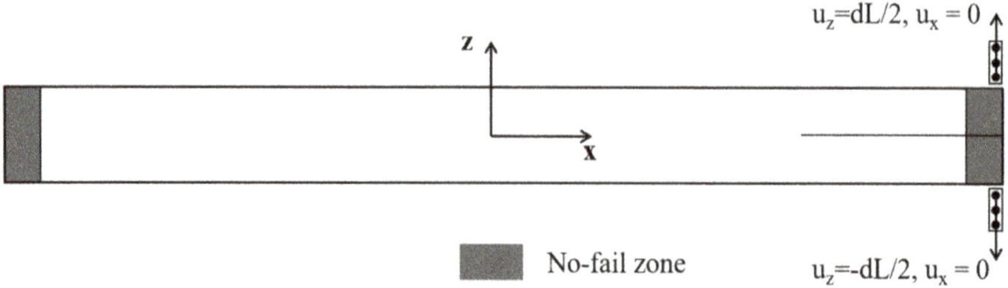

Figure 5. Boundary conditions for PD DCB model.

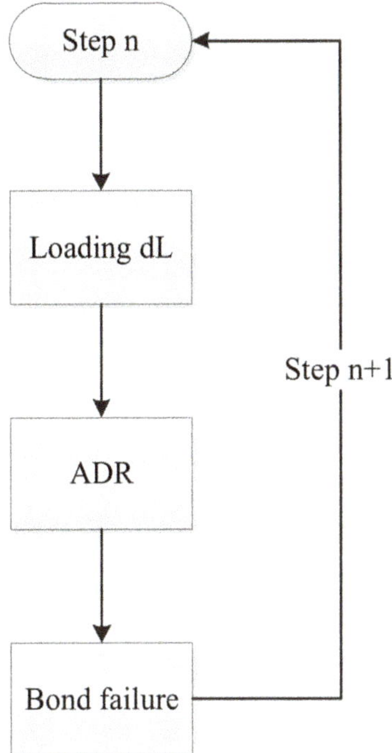

Figure 6. PD DCB modeling flowchart.

4.1. Loading Increment Convergence Analysis and Grid Spacing Influence Study

In order to investigate the reliability of the present modeling, loading increment convergence analysis and grid spacing influence study are conducted. Four different grid spacing dx are studied: 1.0, 0.5, 0.25, and 0.125 (mm). For each grid spacing, four different loading increments dL are studied:

(1) For dx = 1.0, dL: 0.128, 0.064, 0.032, 0.016 (mm);
(2) For dx = 0.5, dL: 0.064, 0.032, 0.016, 0.008 (mm);
(3) For dx = 0.25, dL: 0.032, 0.016, 0.008, 0.004 (mm);
(4) For dx = 0.125, dL: 0.016, 0.008, 0.004, 0.002 (mm).

Firstly, loading increment convergence analysis for each grid spacing dx was conducted, as shown in Figures 7–10. It can be seen that with the decrease of loading increment dL, the PD modeling load–displacement curve converges for all grid spacing. As expected, the slope of the linear load–displacement part is not influenced by the loading increment, and the loading increment only influences the peak force value for a fixed grid spacing.

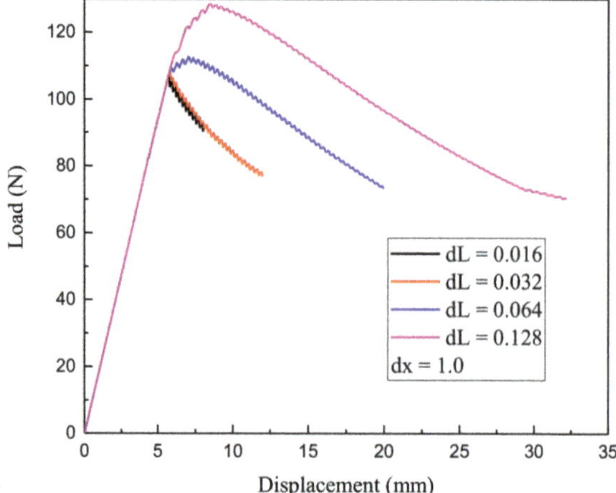

Figure 7. Convergence analysis of loading increment dL for grid spacing dx = 1.0 mm, PD modeling load–displacement.

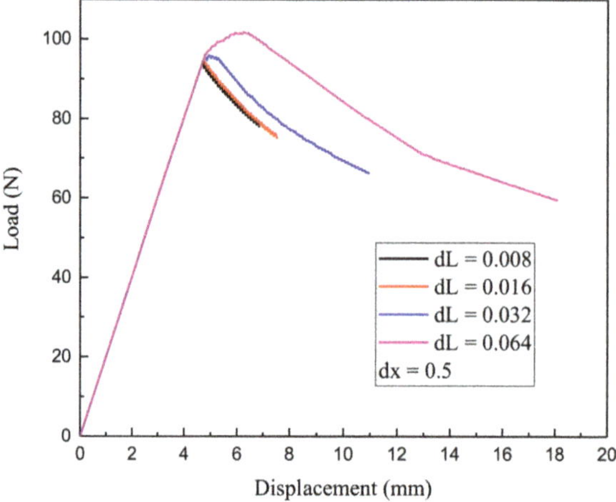

Figure 8. Convergence analysis of loading increment dL for grid spacing dx = 0.5 mm, PD modeling load–displacement.

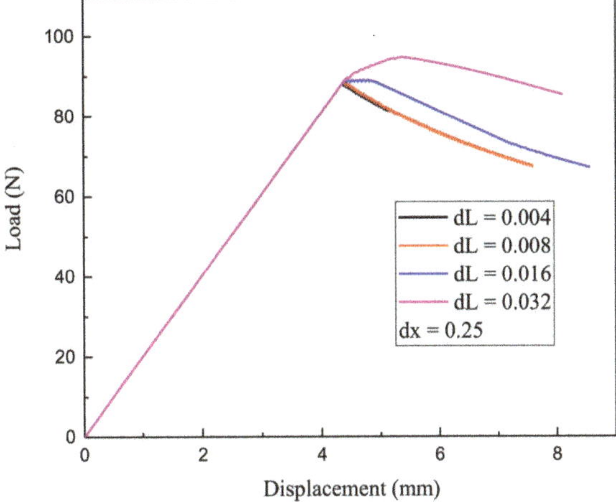

Figure 9. Convergence analysis of loading increment dL for grid spacing dx = 0.25 mm, PD modeling load–displacement.

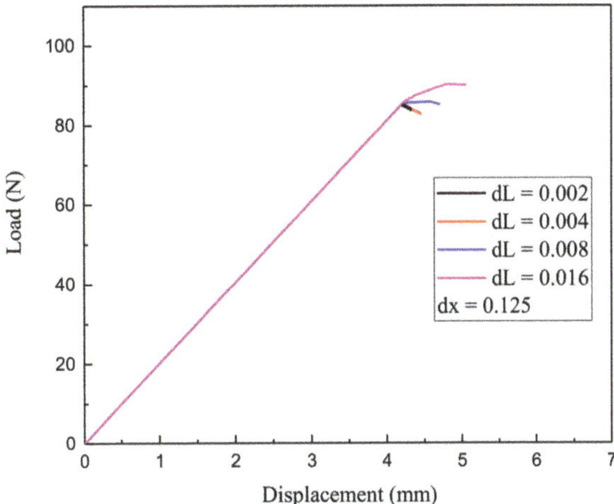

Figure 10. Convergence analysis of loading increment dL for grid spacing dx = 0.125 mm, PD modeling load–displacement.

Then, the influence of grid spacing was studied in Figure 11. The second least loading increment, dL, was used for each grid spacing. For example, dL = 0.008 mm was used for grid spacing dx = 0.25 mm. It can be seen that with the decrease of grid spacing, the linear loading part of the PD modeling load–displacement curve converges. However, PD modeling delamination growth initial point of load–displacement curve seems sensitive to grid spacing. Similar mesh sensitivity of PD modeling can be found in the work by Beckmann et al. [34], Henke et al. [35], and Steward et al. [36]. In the present study, in order to balance the computation reliability and efficiency, dx = 0.25 mm and dL = 0.008 mm are used in the following discussion.

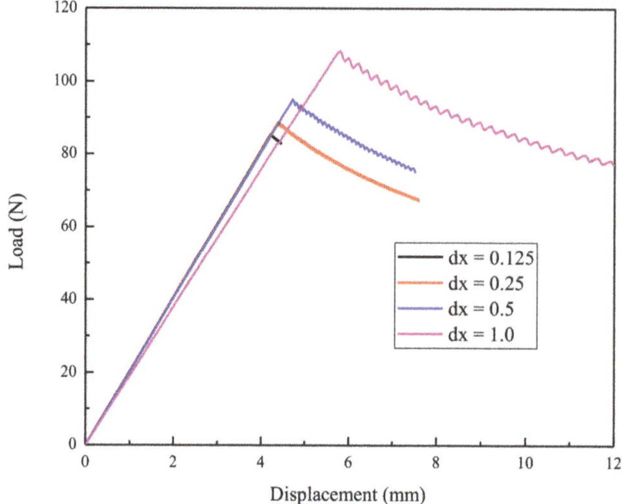

Figure 11. Influence of different grid spacing, PD modeling load–displacement.

4.2. Comparison of Load–Displacement Curve and Delamination Growth Process

The DCB load–displacement curve of the PD and experimental results are compared as shown in Figure 12. The load–displacement curve of PD delamination growth is shown in Figure 13. The results in Figure 12 show that the PD-based analytical result agrees well with experimental results. It is also noted that the modeling load–displacement curve of PD delamination growth shows a zigzag-shaped curve in Figure 13. This is due to the fact that when the delamination front grows for a grid spacing, the displacement-controlled load will drop. Before the next delamination front growth happens, the load will slightly increase with the increase of displacement. A similar phenomenon is also observed in the load–displacement curves obtained from experiments, as shown in Figure 12. However, due to the complex specimen manufacturing process, preset delamination condition, and experimental setup condition, the load–displacement curve obtained from experiments show significant discrepancy between different specimens.

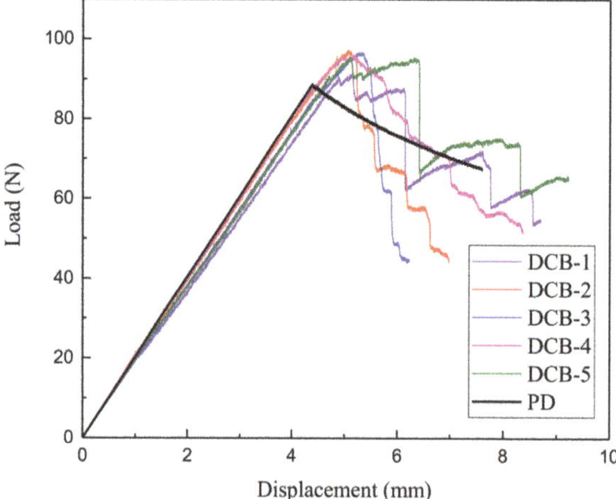

Figure 12. Comparison of load–displacement curve between DCB test and PD (dx = 0.25 mm, dL = 0.008 mm).

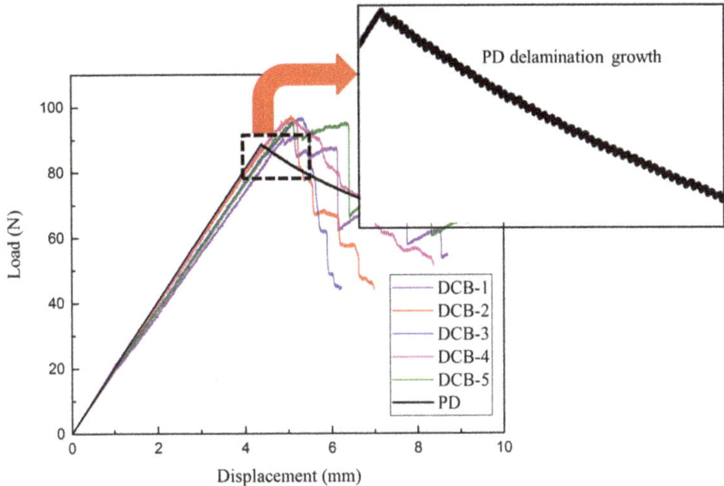

Figure 13. Zigzag-shaped load–displacement curve of DCB test delamination growth process.

Quantitative delamination growth process is also compared between PD and experiment as presented in Table 3, and the delamination growth process damage contours are illustrated in Figure 14. In Table 3, the average delamination growth process of the five DCB test samples from 50 (initial delamination) to 65 mm (shown in Table 1) is compared with the modeling results. From the results, it noted that the present PD simulation results for the composite DCB delamination growth process agree well with the experimental results. The maximum difference between PD and experimental average results in displacement is −14.10%, and is 11.78% in load. In Figure 14, the delamination damage coefficient $\varphi_{\text{delamination}}^{(16)(17)}$ is calculated using Equation (24). The delamination length growing from 50 (initial delamination) to 65 mm is vividly shown. It is worth noting that the PD delamination growth process shown in Figure 14 is "spontaneous". It does not need presetting of delamination path, nor non-physical stabilization parameters which are required for FEM modeling [26]. Also, it does not require the refinement of delamination front grid spacing.

Table 3. Comparison between PD delamination growth with experimental average results (DELL: delamination length).

DELL (mm)	Experimental Avg.		PD		Difference	
	Disp. (mm)	Load (N)	Disp. (mm)	Load (N)	Disp. (mm)	Load (N)
50	5.09	91.35	4.39	88.66	−13.78%	−2.95%
51	5.30	87.66	4.55	87.16	−14.10%	−0.57%
52	5.43	85.07	4.72	85.64	−13.00%	0.67%
53	5.49	85.23	4.89	84.13	−11.00%	−1.30%
54	5.52	84.14	5.06	82.62	−8.29%	−1.81%
55	5.70	81.99	5.23	81.24	−8.26%	−0.91%
60	6.09	67.14	6.16	75.05	1.16%	11.78%
65	6.84	63.64	7.16	69.72	4.66%	9.55%

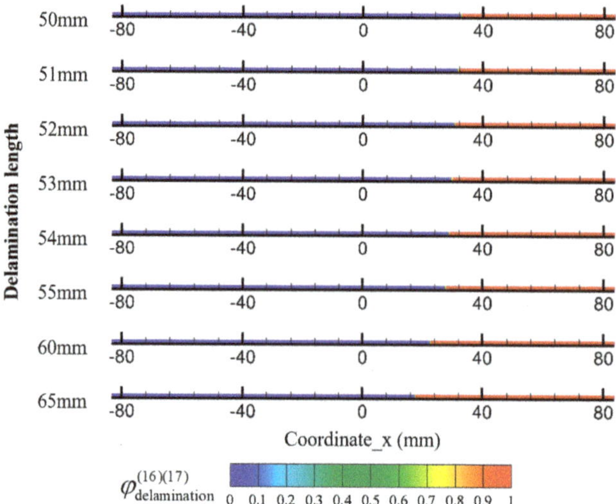

Figure 14. PD modeling delamination growth process damage contours on the mid-plane of DCB specimen, between Layer 16 and 17.

4.3. Comparison of Numerical Modeling Using Revised and Original Energy-Based Failure Criteria

In order to investigate the influence of the revised energy-based failure criteria proposed in Section 2.2, a comparison study is conducted by modeling the DCB delamination using the revised and original energy-based failure criteria. The revised energy-based failure criteria are illustrated in Equation (22). And the original energy-based failure criteria [15] for two-dimensional modeling are

$$\left|\frac{w_\mathrm{I}}{w_\mathrm{IC}}\right| \geq 1, \ w_\mathrm{IC} = \frac{G_\mathrm{IC}}{\frac{4}{3}\delta^3}, \ w_\mathrm{I} = t_z(w' - w). \tag{31}$$

The modeling load–displacement curve using revised and original energy-based failure criteria, as well as the experimental results, are shown in Figure 15. The comparison of maximum load using the revised and original energy-based failure criteria and the average experimental results are presented in Table 4.

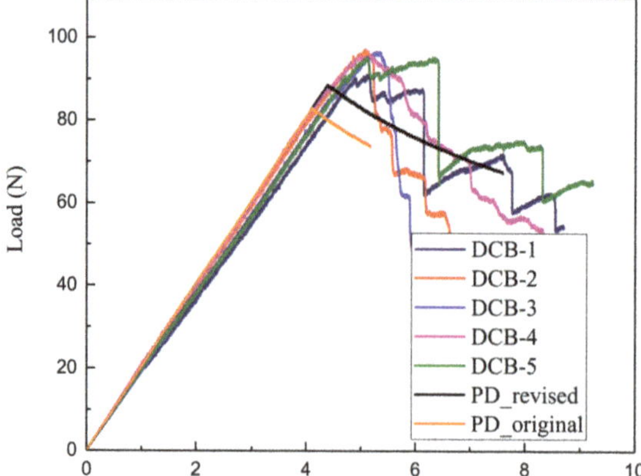

Figure 15. Comparison of load–displacement curve obtained using revised and original bond energy-based failure criteria.

Table 4. Influence of the revised bond energy-based failure criteria on maximum load.

Comparison	Experimental Avg.	PD_original	PD_revised
Max load	95.29	82.97	88.66
Difference	/	−12.93%	−6.96%

From Figure 15, it is observed that the modeling load–displacement curve using revised and original energy-based failure criteria overlapped each other before delamination initiation. However, the modeling delamination initiation was delayed using the revised energy-based failure criteria, which agrees more closely with the experimental results. As shown in Table 4, the differences in the maximum load of the modeling load–displacement curve in comparison with the average experimental results was reduced from −12.93% to −6.96% by using the revised energy-based failure criteria. The above results show that the revised energy-based failure criteria can significantly improve the accuracy of the PD delamination modeling of DCB test.

5. Conclusions

In this study, a two-dimensional ordinary state-based peridynamic (OSB PD) modeling of mode-I delamination growth in double cantilever composite beam (DCB) test is conducted. The two-dimensional OSB PD composite model for DCB modeling is obtained by reformulating the previous OSB PD lamina model in x–z direction. Additionally, revised energy-based failure criteria are proposed for modeling delamination initiation and growth.

In order to investigate the reliability of current modeling, loading increment convergence analysis and grid spacing influence study are conducted. It is shown that the PD modeling load–displacement curve converges with the decrease of loading increment. The linear loading part of PD modeling load–displacement curve converges with the decrease of grid spacing. However, the delamination growth initial point of PD modeling load–displacement curve seems sensitive to grid spacing, similar to the previous literature work.

The standard DCB test was performed with five composite DCB specimens to measure the mode-I delamination fracture toughness G_{IC}. The load–displacement variation and delamination growths were also measured. The modeling load–displacement curve and delamination growth process using the present PD model are then compared with experimental results. The results show that the PD-based load–displacement curve agrees well with experimental results. The zigzag load–displacement curve for delamination growth was observed both in experimental results and the present PD modeling results. Well agreement of the PD modeling delamination growth progress with experimental results is achieved. In particular, the maximum difference between PD and experimental average results in displacement is −14.10%, and in load is 11.78%. Besides, the PD modeling delamination growth process damage contours are illustrated.

To demonstrate the influence of the revised energy-based failure criteria, a comparison study using the revised and original energy-based failure criteria in the PD modeling is conducted. It is shown that the modeling delamination initiation was delayed using the revised energy-based failure criteria, which agrees more closely with the experimental results. In particular, by using the revised energy-based failure criteria, the differences with experimental results in the maximum load of the modeling load–displacement curve was reduced from −12.93% to −6.96%, which show significant improvement in the accuracy of delamination modeling of DCB test.

Author Contributions: Supervision, S.G. and H.W.; Writing—original draft preparation, X.-W.J.; Writing—review and editing, H.L.

Funding: This research was funded by China Scholarship Council (CSC NO. 201706230169).

Acknowledgments: The authors acknowledge the financial support from China Scholarship Council (CSC NO. 201706230169).

Conflicts of Interest: The authors declare no conflict of interest.

Appendix A

Appendix A.1. Stiffness Matrix of DCB Composite Specimen in x–z Direction

Assuming plane strain condition for DCB composite specimen (shown in Figures 4 and 5) in x–z direction:

$$\varepsilon_{22} = 0, \ \gamma_{12} = \gamma_{23} = 0, \ \tau_{12} = \tau_{23} = 0. \tag{A1}$$

Similar to plane stress assumption for unidirectional lamina in x–y direction, the strain–stress relationship and compliance matrix can be written as

$$\begin{bmatrix} \varepsilon_1 \\ \varepsilon_3 \\ \gamma_{13} \end{bmatrix} = \begin{bmatrix} S_{11} & S_{13} & 0 \\ S_{13} & S_{33} & 0 \\ 0 & 0 & S_{55} \end{bmatrix} \begin{bmatrix} \sigma_1 \\ \sigma_3 \\ \tau_{13} \end{bmatrix}. \tag{A2}$$

Using uniaxial loading condition, the compliance matrix can be related to the engineering elastic constants of CFRP composite:

(1) $\sigma_1 \neq 0, \ \sigma_3 = 0, \ \tau_{13} = 0$

$$\varepsilon_1 = S_{11}\sigma_1, \ \varepsilon_3 = S_{13}\sigma_1, \ \gamma_{13} = 0, \tag{A3}$$

$$E_1 = \frac{\sigma_1}{\varepsilon_1} = \frac{\sigma_1}{S_{11}\sigma_1} = \frac{1}{S_{11}}, \tag{A4}$$

$$\nu_{13} = -\frac{\varepsilon_3}{\varepsilon_1} = -\frac{S_{13}\sigma_1}{S_{11}\sigma_1} = -\frac{S_{13}}{S_{11}}, \tag{A5}$$

$$S_{11} = \frac{1}{E_1}, \tag{A6}$$

$$S_{13} = -S_{11}\nu_{13} = -\frac{\nu_{13}}{E_1}. \tag{A7}$$

(2) $\sigma_3 \neq 0, \ \sigma_1 = 0, \ \tau_{13} = 0$

$$\varepsilon_1 = S_{13}\sigma_3, \ \varepsilon_3 = S_{33}\sigma_3, \ \gamma_{13} = 0, \tag{A8}$$

$$E_3 = \frac{\sigma_3}{\varepsilon_3} = \frac{\sigma_3}{S_{33}\sigma_3} = \frac{1}{S_{33}}, \tag{A9}$$

$$\nu_{31} = -\frac{\varepsilon_1}{\varepsilon_3} = -\frac{S_{13}\sigma_3}{S_{33}\sigma_3} = -\frac{S_{13}}{S_{33}}, \tag{A10}$$

$$S_{33} = \frac{1}{E_3}. \tag{A11}$$

(3) $\tau_{13} \neq 0, \sigma_1 = 0, \ \sigma_3 = 0$

$$\varepsilon_1 = 0, \ \varepsilon_3 = 0, \ \gamma_{13} = S_{55}\tau_{13}, \tag{A12}$$

$$G_{13} = \frac{\tau_{13}}{\gamma_{13}} = \frac{\tau_{13}}{S_{55}\tau_{13}} = \frac{1}{S_{55}}, \tag{A13}$$

$$S_{55} = \frac{1}{G_{13}}. \tag{A14}$$

The stiffness matrix of DCB composite specimen in x–z direction can be proposed by the inverse of compliance matrix,

$$\begin{bmatrix} \sigma_1 \\ \sigma_3 \\ \tau_{13} \end{bmatrix} = \begin{bmatrix} Q_{11} & Q_{13} & 0 \\ Q_{13} & Q_{33} & 0 \\ 0 & 0 & Q_{55} \end{bmatrix} \begin{bmatrix} \varepsilon_1 \\ \varepsilon_3 \\ \gamma_{13} \end{bmatrix}, \tag{A15}$$

$$[Q] = [S]^{-1} = \begin{bmatrix} \dfrac{S_{33}}{S_{11}S_{33}-S_{13}{}^2} & -\dfrac{S_{13}}{S_{11}S_{33}-S_{13}{}^2} & 0 \\ -\dfrac{S_{13}}{S_{11}S_{33}-S_{13}{}^2} & \dfrac{S_{11}}{S_{11}S_{33}-S_{13}{}^2} & 0 \\ 0 & 0 & \dfrac{1}{S_{55}} \end{bmatrix}, \tag{A16}$$

$$Q_{11} = \frac{S_{33}}{S_{11}S_{33} - S_{13}{}^2} = \frac{E_1}{1 - \nu_{13}\nu_{31}}, \tag{A17}$$

$$Q_{13} = -\frac{S_{13}}{S_{11}S_{33} - S_{13}{}^2} = \frac{\nu_{13}E_3}{1 - \nu_{13}\nu_{31}}, \tag{A18}$$

$$Q_{33} = \frac{S_{11}}{S_{11}S_{33} - S_{13}{}^2} = \frac{E_3}{1 - \nu_{13}\nu_{31}}, \tag{A19}$$

$$Q_{55} = \frac{1}{S_{55}} = G_{13}. \tag{A20}$$

Appendix A.2. Derivation Procedure of PD Material Parameters

The PD strain energy density of the present two-dimensional OSB PD composite model for CFRP composite can be expressed as

$$W_{(k)} = a\theta_{(k)}^2 + b\sum_{j=1}^{\infty} \delta \left| x_{(j)} - x_{(k)} \right| s_{(k)(j)}^2 V_{(j)} + b_F \sum_{j=1}^{J} \delta \left| x_{(j)} - x_{(k)} \right| s_{(k)(j)}^2 V_{(j)} + b_Z \sum_{j=1}^{J} \delta \left| x_{(j)} - x_{(k)} \right| s_{(k)(j)}^2 V_{(j)}. \tag{A21}$$

The PD material parameters in Equations (4), (5) and (10) can be derived by comparing the PD strain energy density and the strain energy density of continuum mechanics under simple loading conditions. Four simple loading conditions are assumed: (1) Pure shear loading on x–z plane (γ_{13}); (2) Uniaxial stretch in fiber direction (ε_{11}); (3) Uniaxial stretch in thickness direction (ε_{33}); (4) Biaxial tension on x–z plane (ε_{11}, ε_{33}).

(1) Pure shear loading on x–z plane (γ_{13})

Under this loading condition, we assume $\gamma_{13} = \zeta$ and all other strains equal zero. From the derivation in Appendix A.1,

$$\theta_{(k)}^{CM} = 0, \ W_{(k)}^{CM} = \frac{1}{2}Q_{55}\zeta^2. \tag{A22}$$

Firstly, the stretch of bonds $s_{(k)(j)}$ should be calculated under $\gamma_{13} = \zeta$. As shown in Figure A1, under this simple loading condition, material point M moves to M'.

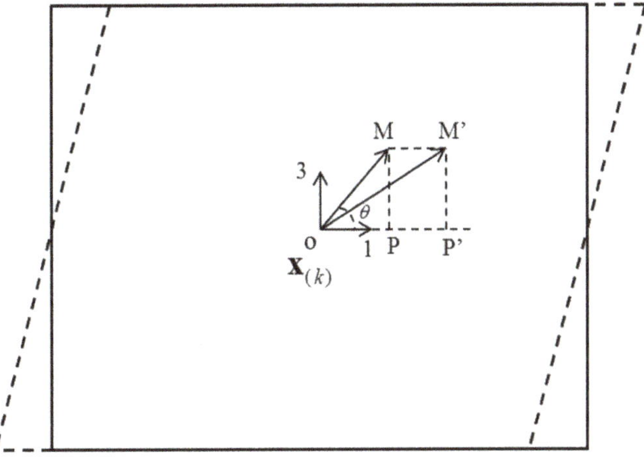

Figure A1. Pure shear loading on x–z plane (γ_{13}).

$$
\begin{aligned}
OM' &= \sqrt{OP'^2 + M'P'^2} = \sqrt{(\xi\cos\theta + \xi\sin\theta\zeta)^2 + (\xi\sin\theta)^2} \\
&= \sqrt{\xi^2 + 2\xi^2\cos\theta\sin\theta\zeta + \xi^2\sin^2\theta\zeta^2} \\
&\approx \sqrt{\xi^2 + 2\xi^2\cos\theta\sin\theta\zeta}
\end{aligned}
\tag{A23}
$$

$$
\begin{aligned}
s_{(k)(j)} &= \frac{OM' - OM}{OM} = \frac{\sqrt{\xi^2 + 2\xi^2\cos\theta\sin\theta\zeta} - \xi}{\xi} = \sqrt{1 + \zeta 2\cos\theta\sin\theta} - 1 \\
&\approx 1 + \tfrac{1}{2}\zeta 2\cos\theta\sin\theta - 1 = \zeta\sin\theta\cos\theta
\end{aligned}
\tag{A24}
$$

Then the PD dilatation can be derived as

$$
\begin{aligned}
\theta_{(k)} &= d\sum_{j=1}^{\infty}\delta s_{(k)(j)}\Lambda_{(k)(j)}V_{(j)} = d\sum_{j=1}^{\infty}\delta\zeta\sin\theta\cos\theta\Lambda_{(k)(j)}V_{(j)} \\
&= d\int_0^{\delta}\int_0^{2\pi}\delta\zeta\sin\theta\cos\theta\xi\,d\theta\,d\xi = 0
\end{aligned}
\tag{A25}
$$

and the PD strain energy could be derived as

$$
\begin{aligned}
W_{(k)} &= a\theta_{(k)}^2 + b\sum_{j=1}^{\infty}\delta\big|x_{(j)} - x_{(k)}\big|s_{(k)(j)}^2 V_{(j)} + b_F\sum_{j=1}^{J}\delta\big|x_{(j)} - x_{(k)}\big|s_{(k)(j)}^2 V_{(j)} + b_Z\sum_{j=1}^{J}\delta\big|x_{(j)} - x_{(k)}\big|s_{(k)(j)}^2 V_{(j)} \\
&= a(0) + b\int_H \delta\xi(\zeta\sin\theta\cos\theta)^2 dH + b_F(0) + b_Z(0) \\
&= b\int_0^{\delta}\int_0^{2\pi}\delta\xi(\zeta\sin\theta\cos\theta)^2\xi\,d\theta\,d\xi = \frac{\pi\delta^4\zeta^2}{12}b = \tfrac{1}{2}Q_{55}\zeta^2
\end{aligned}
\tag{A26}
$$

Comparing with Equation (A22),

$$
b = \frac{\tfrac{1}{2}Q_{55}\zeta^2}{\frac{\pi\delta^4\zeta^2}{12}} = \frac{6Q_{55}}{\pi\delta^4}.
\tag{A27}
$$

(2) Uniaxial stretch in fiber direction (ε_{11})

Setting $\varepsilon_{11} = \zeta$, similarly,

$$
\theta_{(k)}^{CM} = \zeta, \quad W_{(k)}^{CM} = \frac{1}{2}Q_{11}\zeta^2,
\tag{A28}
$$

$$
s_{(k)(j)} = \zeta\cos^2\theta,
\tag{A29}
$$

$$
\begin{aligned}
\theta_{(k)} &= d\sum_{j=1}^{\infty}\delta s_{(k)(j)}\Lambda_{(k)(j)}V_{(j)} = d\sum_{j=1}^{\infty}\delta\zeta\cos^2\theta\Lambda_{(k)(j)}V_{(j)} \\
&= d\int_0^{\delta}\int_0^{2\pi}\delta\zeta\cos^2\theta\xi\,d\theta\,d\xi = \frac{\pi}{2}d\delta^3\zeta = \zeta
\end{aligned}
\tag{A30}
$$

$$
\begin{aligned}
W_{(k)} &= a\theta_{(k)}^2 + b\sum_{j=1}^{\infty}\delta|x_{(j)} - x_{(k)}|s_{(k)(j)}^2 V_{(j)} + b_F\sum_{j=1}^{J}\delta|x_{(j)} - x_{(k)}|s_{(k)(j)}^2 V_{(j)} + b_Z\sum_{j=1}^{J}\delta|x_{(j)} - x_{(k)}|s_{(k)(j)}^2 V_{(j)} \\
&= a\zeta^2 + b\int_H \delta\xi(\zeta\cos^2\theta)^2 dH + b_F\sum_{j=1}^{J}\delta|\mathbf{x}_{(j)} - \mathbf{x}_{(k)}|(\zeta\cos^2\theta)^2 V_{(j)} + b_Z(0) \\
&= a\zeta^2 + b\int_0^{\delta}\int_0^{2\pi}\delta\xi(\zeta\cos^2\theta)^2\xi\,d\theta\,d\xi + \zeta^2 b_F\delta\sum_{j=1}^{J}|\mathbf{x}_{(j)} - \mathbf{x}_{(k)}|V_{(j)} \\
&= a\zeta^2 + \tfrac{\pi}{4}b\delta^4\zeta^2 + \zeta^2 b_F\delta\sum_{j=1}^{J}|\mathbf{x}_{(j)} - \mathbf{x}_{(k)}|V_{(j)} = \tfrac{1}{2}Q_{11}\zeta^2
\end{aligned}
\tag{A31}
$$

Comparing Equation (A28) with (A30) and (A31),

$$
d = \frac{2}{\pi\delta^3},
\tag{A32}
$$

$$
a + \frac{\pi}{4}b\delta^4 + b_F\delta\sum_{j=1}^{J}\big|x_{(j)} - x_{(k)}\big|V_{(j)} = \frac{1}{2}Q_{11}.
\tag{A33}
$$

Substituting Equation (A27),

$$a + b_F \delta \sum_{j=1}^{J} |x_{(j)} - x_{(k)}| V_{(j)} = \frac{1}{2} Q_{11} - \frac{3}{2} Q_{55}. \tag{A34}$$

(3) Uniaxial stretch in thickness direction (ε_{33})

Setting $\varepsilon_{33} = \zeta$, similarly,

$$\theta_{(k)}^{CM} = \zeta, \ W_{(k)}^{CM} = \frac{1}{2} Q_{33} \zeta^2, \tag{A35}$$

$$s_{(k)(j)} = \zeta \sin^2 \theta, \tag{A36}$$

$$\begin{aligned}
\theta_{(k)} &= d \sum_{j=1}^{\infty} \delta s_{(k)(j)} \Lambda_{(k)(j)} V_{(j)} = d \sum_{j=1}^{\infty} \delta \zeta \sin^2 \theta \Lambda_{(k)(j)} V_{(j)} \\
&= d \int_0^\delta \int_0^{2\pi} \delta \zeta \sin^2 \theta \xi d\theta d\xi = \frac{\pi}{2} d \delta^3 \zeta = \zeta
\end{aligned} \tag{A37}$$

$$\begin{aligned}
W_{(k)} &= a\theta_{(k)}^2 + b \sum_{j=1}^{\infty} \delta |x_{(j)} - x_{(k)}| s_{(k)(j)}^2 V_{(j)} + b_F \sum_{j=1}^{J} \delta |x_{(j)} - x_{(k)}| s_{(k)(j)}^2 V_{(j)} + b_Z \sum_{j=1}^{J} \delta |x_{(j)} - x_{(k)}| s_{(k)(j)}^2 V_{(j)} \\
&= a\zeta^2 + b \int_H \delta \xi (\zeta \sin^2 \theta)^2 dH + b_F(0) + b_Z \sum_{j=1}^{J} \delta |\mathbf{x}_{(j)} - \mathbf{x}_{(k)}| (\zeta \sin^2 \theta)^2 V_{(j)} \\
&= a\zeta^2 + b \int_0^\delta \int_0^{2\pi} \delta \xi (\zeta \sin^2 \theta)^2 \xi d\theta d\xi + \zeta^2 b_Z \delta \sum_{j=1}^{J} |\mathbf{x}_{(j)} - \mathbf{x}_{(k)}| V_{(j)} \\
&= a\zeta^2 + \frac{\pi}{4} b \delta^4 \zeta^2 + \zeta^2 b_Z \delta \sum_{j=1}^{J} |\mathbf{x}_{(j)} - \mathbf{x}_{(k)}| V_{(j)} = \frac{1}{2} Q_{33} \zeta^2
\end{aligned} \tag{A38}$$

Comparing Equation (A35) with (A37) and (A38),

$$d = \frac{2}{\pi \delta^3}, \tag{A39}$$

$$a + \frac{\pi}{4} b \delta^4 + b_Z \delta \sum_{j=1}^{J} \left| x_{(j)} - x_{(k)} \right| V_{(j)} = \frac{1}{2} Q_{33}. \tag{A40}$$

Substituting Equation (A27),

$$a + b_Z \delta \sum_{j=1}^{J} \left| x_{(j)} - x_{(k)} \right| V_{(j)} = \frac{1}{2} Q_{33} - \frac{3}{2} Q_{55}. \tag{A41}$$

(4) Biaxial tension on x–z plane (ε_{11}, ε_{33})

Setting $\varepsilon_{11} = \zeta$, $\varepsilon_{33} = \zeta$, similarly,

$$\theta_{(k)}^{CM} = 2\zeta, \ W_{(k)}^{CM} = \frac{1}{2}(Q_{11} + 2Q_{13} + Q_{33})\zeta^2, \tag{A42}$$

$$s_{(k)(j)} = \zeta, \tag{A43}$$

$$\begin{aligned}
\theta_{(k)} &= d \sum_{j=1}^{\infty} \delta s_{(k)(j)} \Lambda_{(k)(j)} V_{(j)} = d \sum_{j=1}^{\infty} \delta \zeta \Lambda_{(k)(j)} V_{(j)} \\
&= d \int_0^\delta \int_0^{2\pi} \delta \zeta \xi d\theta d\xi = d \delta^3 \pi \zeta = 2\zeta
\end{aligned} \tag{A44}$$

$$
\begin{aligned}
W_{(k)} \;&= a\theta_{(k)}^2 + b\sum_{j=1}^{\infty}\delta|x_{(j)}-x_{(k)}|s_{(k)(j)}^2 V_{(j)} + b_F\sum_{j=1}^{J}\delta|x_{(j)}-x_{(k)}|s_{(k)(j)}^2 V_{(j)} + b_Z\sum_{j=1}^{J}\delta|x_{(j)}-x_{(k)}|s_{(k)(j)}^2 V_{(j)} \\[4pt]
&= a(2\zeta)^2 + b\int_H \delta\xi\zeta^2 dH + b_F\sum_{j=1}^{J}\delta|\mathbf{x}_{(j)}-\mathbf{x}_{(k)}|\zeta^2 V_{(j)}\, b_Z\sum_{j=1}^{J}\delta|\mathbf{x}_{(j)}-\mathbf{x}_{(k)}|\zeta^2 V_{(j)} \\[4pt]
&= 4a\zeta^2 + b\int_0^{\delta}\int_0^{2\pi}\delta\xi\zeta^2\xi d\theta d\xi + \zeta^2 b_F\delta\sum_{j=1}^{J}|\mathbf{x}_{(j)}-\mathbf{x}_{(k)}|V_{(j)}\zeta^2 b_Z\delta\sum_{j=1}^{J}|\mathbf{x}_{(j)}-\mathbf{x}_{(k)}|V_{(j)} \\[4pt]
&= 4a\zeta^2 + \tfrac{2}{3}\pi b\delta^4\zeta^2 + \zeta^2 b_F\delta\sum_{j=1}^{J}|\mathbf{x}_{(j)}-\mathbf{x}_{(k)}|V_{(j)} + \zeta^2 b_Z\delta\sum_{j=1}^{J}|\mathbf{x}_{(j)}-\mathbf{x}_{(k)}|V_{(j)} = \tfrac{1}{2}Q_{11}+2Q_{13}+Q_{33})\zeta^2
\end{aligned}
\tag{A45}
$$

Comparing Equation (A42) with (A44) and (A45),

$$
d = \frac{2}{\pi\delta^3}, \tag{A46}
$$

$$
4a + b_F\delta\sum_{j=1}^{J}\left|x_{(j)}-x_{(k)}\right|V_{(j)} + b_Z\delta\sum_{j=1}^{J}\left|x_{(j)}-x_{(k)}\right|V_{(j)} = \frac{1}{2}(Q_{11}+2Q_{13}+Q_{33}-8Q_{55}). \tag{A47}
$$

Solving Equations (A34), (A41), and (A47),

$$
a = \frac{1}{2}(Q_{13}-Q_{55}), \tag{A48}
$$

$$
b_F = \frac{Q_{11}-Q_{13}-2Q_{55}}{2\delta\sum\limits_{j=1}^{J}\left|x_{(j)}-x_{(k)}\right|V_{(j)}}, \quad b_Z = \frac{Q_{33}-Q_{13}-2Q_{55}}{2\delta\sum\limits_{j=1}^{J}\left|x_{(j)}-x_{(k)}\right|V_{(j)}}. \tag{A49}
$$

References

1. Alfano, G.; Crisfield, M.A. Finite element interface models for the delamination analysis of laminated composites: Mechanical and computational issues. *Int. J. Numer. Methods Eng.* **2001**, *50*, 1701–1736. [CrossRef]
2. Camanho, P.P.; Vila, C.G.D.; Ambur, D.R. *Numerical Simulation of Delamination Growth in Composite Materials*; NASA Technical Paper 211041; NASA: Washington, DC, USA, 2001; pp. 1–24.
3. Camanho, P.P.; Dávila, C.G.; De Moura, M.F. Numerical simulation of mixed-mode progressive delamination in composite materials. *J. Compos. Mater.* **2003**, *37*, 1415–1438. [CrossRef]
4. Meo, M.; Thieulot, E. Delamination modelling in a double cantilever beam. *Compos. Struct.* **2005**, *71*, 429–434. [CrossRef]
5. Zhao, L.; Zhi, J.; Zhang, J.; Liu, Z.; Hu, N. XFEM simulation of delamination in composite laminates. *Compos. Part A Appl. Sci. Manuf.* **2016**, *80*, 61–71. [CrossRef]
6. Kim, S.H.; Kim, C.G. Optimal design of composite stiffened panel with cohesive elements using micro-genetic algorithm. *J. Compos. Mater.* **2008**, *42*, 2259–2273. [CrossRef]
7. Orifici, A.C.; Shah, S.A.; Herszberg, I.; Kotler, A.; Weller, T. Failure analysis in postbuckled composite T-sections. *Compos. Struct.* **2008**, *86*, 146–153. [CrossRef]
8. Riccio, A.; Raimondo, A.; Di Felice, G.; Scaramuzzino, F. A numerical procedure for the simulation of skin-stringer debonding growth in stiffened composite panels. *Aerosp. Sci. Technol.* **2014**, *39*, 307–314. [CrossRef]
9. Oterkus, E.; Madenci, E. Peridynamic Theory for Damage Initiation and Growth in Composite Laminate. *Key Eng. Mater.* **2012**, *488–489*, 355–358. [CrossRef]
10. Xu, J.; Askari, A.; Weckner, O.; Silling, S. Peridynamic Analysis of Impact Damage in Composite Laminates. *J. Aerosp. Eng.* **2008**, *21*, 187–194. [CrossRef]
11. Askari, E.; Xu, J.; Silling, S. Peridynamic Analysis of Damage and Failure in Composites. In Proceedings of the 44th AIAA Aerospace Sciences Meeting and Exhibit, Reno, NV, USA, 9–12 January 2006; American Institute of Aeronautics and Astronautics: Reston, VA, USA, 2006; pp. 1–12.

12. Hu, Y.L.; De Carvalho, N.V.; Madenci, E. Peridynamic modeling of delamination growth in composite laminates. *Compos. Struct.* **2015**, *132*, 610–620. [CrossRef]

13. Silling, S.A.; Epton, M.; Weckner, O.; Xu, J.; Askari, E. Peridynamic states and constitutive modeling. *J. Elast.* **2007**, *88*, 151–184. [CrossRef]

14. Silling, S.A. Reformulation of elasticity theory for discontinuities and long-range forces. *J. Mech. Phys. Solids* **2000**, *48*, 175–209. [CrossRef]

15. Silling, S.A.; Lehoucq, R.B. Peridynamic Theory of Solid Mechanics. In *Advances in Applied Mechanics*; Aref, H., VanDerGiessen, E., Eds.; Elsevier: Amsterdam, The Netherlands, 2010; pp. 73–168.

16. Diyaroglu, C.; Oterkus, E.; Madenci, E.; Rabczuk, T.; Siddiq, A. Peridynamic modeling of composite laminates under explosive loading. *Compos. Struct.* **2016**, *144*, 14–23. [CrossRef]

17. Oterkus, E.; Madenci, E. Peridynamic Analysis of Fiber Reinforced Composite Materials. *J. Mech. Mater. Struct.* **2012**, *7*, 45–84. [CrossRef]

18. Oterkus, E.; Madenci, E.; Weckner, O.; Silling, S.; Bogert, P.; Tessler, A. Combined finite element and peridynamic analyses for predicting failure in a stiffened composite curved panel with a central slot. *Compos. Struct.* **2012**, *94*, 839–850. [CrossRef]

19. Oterkus, E.; Barut, A.; Madenci, E. Damage Growth Prediction from Loaded Composite Fastener Holes by Using Peridynamic Theory. In Proceedings of the 51th AIAA/ASME/ASCE/AHS/ASC Structures, Structural Dynamics, and Materials Conference, Orlando, FL, USA, 12–15 April 2010. Paper No.2010-3026.

20. Hu, Y.; Yu, Y.; Wang, H. Peridynamic analytical method for progressive damage in notched composite laminates. *Compos. Struct.* **2014**, *108*, 801–810. [CrossRef]

21. Hu, Y.L.; Madenci, E. Bond-based peridynamic modeling of composite laminates with arbitrary fiber orientation and stacking sequence. *Compos. Struct.* **2016**, *153*, 139–175. [CrossRef]

22. Hu, Y.; Madenci, E.; Phan, N.D. Peridynamics for Predicting Tensile and Compressive Strength of Notched Composites. In Proceedings of the 57th AIAA/ASCE/AHS/ASC Structures, Structural Dynamics, and Materials Conference, San Diego, CA, USA, 4–8 January 2016; American Institute of Aeronautics and Astronautics: Reston, VA, USA, 2016.

23. Hu, Y.L.; Madenci, E. Peridynamics for fatigue life and residual strength prediction of composite laminates. *Compos. Struct.* **2017**, *160*, 169–184. [CrossRef]

24. Heidari-Rarani, M.; Shokrieh, M.M.; Camanho, P.P. Finite element modeling of mode I delamination growth in laminated DCB specimens with R-curve effects. *Compos. Part B Eng.* **2013**, *45*, 897–903. [CrossRef]

25. Shanmugam, V.; Penmetsa, R.; Tuegel, E.; Clay, S. Stochastic modeling of delamination growth in unidirectional composite DCB specimens using cohesive zone models. *Compos. Struct.* **2013**, *102*, 38–60. [CrossRef]

26. Krueger, R. A summary of benchmark examples to assess the performance of quasi-static delamination propagation prediction capabilities in finite element codes. *J. Compos. Mater.* **2015**, *49*, 3297–3316. [CrossRef]

27. Madenci, E.; Oterkus, E. *Peridynamic Theory and Its Applications*; Springer: New York, NY, USA, 2014.

28. Silling, S.A.; Askari, E. A meshfree method based on the peridynamic model of solid mechanics. *Comput. Struct.* **2005**, *83*, 1526–1535. [CrossRef]

29. Kilic, B.; Madenci, E. An adaptive dynamic relaxation method for quasi-static simulations using the peridynamic theory. *Theor. Appl. Fract. Mech.* **2010**, *53*, 194–204. [CrossRef]

30. Breitenfeld, M.S.; Geubelle, P.H.; Weckner, O.; Silling, S.A. Non-ordinary state-based peridynamic analysis of stationary crack problems. *Comput. Methods Appl. Mech. Eng.* **2014**, *272*, 233–250. [CrossRef]

31. Ni, T.; Zaccariotto, M.; Zhu, Q.-Z.; Galvanetto, U. Static solution of crack propagation problems in Peridynamics. *Comput. Methods Appl. Mech. Eng.* **2019**, *346*, 126–151. [CrossRef]

32. Fatica, M.; Ruetsch, G. *CUDA Fortran for Scientists and Engineers: Best Practices for Efficient CUDA Fortran Programming*; Elsevier: Amsterdam, The Netherlands, 2013.

33. *ASTM: D5528-13: Standard Test Method for Mode I Interlaminar Fracture Toughness of Unidirectional Fiber-Reinforced Polymer Matrix Composites*; American Society for Testing and Materials: West Conshohocken, PA, USA, 2013.

34. Beckmann, R.; Mella, R.; Wenman, M.R. Mesh and timestep sensitivity of fracture from thermal strains using peridynamics implemented in Abaqus. *Comput. Methods Appl. Mech. Eng.* **2013**, *263*, 71–80. [CrossRef]

newpage

35. Henke, S.F.; Shanbhag, S. Mesh sensitivity in peridynamic simulations. *Comput. Phys. Commun.* **2014**, *185*, 181–193. [CrossRef]
36. Steward, R.J.; Jeon, B. Decoupling strength and grid resolution. In Proceedings of the 18th U.S. National Congress for Theoretical and Applied Mechanics, Chicago, IL, USA, 4–9 June 2018.

 © 2019 by the authors. Licensee MDPI, Basel, Switzerland. This article is an open access article distributed under the terms and conditions of the Creative Commons Attribution (CC BY) license (http://creativecommons.org/licenses/by/4.0/).

 *applied sciences*

Article

Analytical Model of Wellbore Stability of Fractured Coal Seam Considering the Effect of Cleat Filler and Analysis of Influencing Factors

Xinbo Zhao [1], Jianjun Wang [2] and Yue Mei [3,4,*]

[1] School of Science, Qingdao University of Technology, Qingdao 266033, China; zxbups@163.com
[2] State Key Laboratory for Performance and Structure Safety of Petroleum Tubular Goods and Equipment Materials, CNPC TubularGoods Research Center, Xian 710077, China; wg_j_jun@163.com
[3] State Key Laboratory of Structural Analysis for Industrial Equipment, Department of Engineering Mechanics, Dalian University of Technology, Dalian 116023, China
[4] International Research Center for Computational Mechanics, Dalian University of Technology, Dalian 116023, China
* Correspondence: meiyue1989@gmail.com or meiyue@dlut.edu.cn

Received: 23 December 2019; Accepted: 30 January 2020; Published: 9 February 2020

Abstract: Currently, coal borehole collapses frequently occur during drilling. Considering that the coal near to the wellbore is cut into blocks, and the cleat filler of the coal influences the stress distribution near the wellbore, a new theoretical solution of a near-wellbore Stress Field in coal bed wells is established. In addition, according to the limit equilibrium theory and the E.MG-C criterion, the limit sliding formula of the quadrilateral and triangular block is deduced, and the slipping direction of the blocks is further judged. Finally, the wellbore stability model of the coal seam is established. The accuracy of the theoretical model is verified through a numerical method by using the PFC software. Based upon this wellbore stability theoretical model of coal, many cleat affecting factors such as cleat spacing, cleat length, cleat angle and the cleat geometric position, are studied, and the results show that a quadrilateral block slides off more easily than a triangular block under the same boundary condition; the bigger the cleat spacing and cleat length are, the lower is the risk that blocks slide off, and increasing the cleat angle could cause blocks to slide off easily. Under the same boundary condition, whether blocks slide off or not is closely related to the well round angle.

Keywords: cleat filler; broken coal seam; wellbore stability; analytical model; affecting factors

---

1. Introduction

It is an essential part of the unconventional development of energy to exploit coalbed methane in China [1]. Under the condition of natural strata, the coalbed presents its discontinuous and fragile properties, due to the limitation of the structural characteristics, such as natural fracture extension, abundant cleat system, low elastic modulus and strong anisotropy [2,3]. When the drilling contacts with this kind of broken coal seam, it tends to cause the collapse of the coal wellbore by the disturbance of the drilling string and the intrusion of the drilling fluid [4–6]. Until now, many specialists have carried out a number of studies about the stability of the sandstone wall through the continuum theory, which is useful for quasi-continuous reservoirs, such as sandstone, but is not useful for a broken coal seam. So, it is urgent to study the mechanism of the wellbore instability of the broken coal seam.

Specialists have achieved a number of results [7–10] on the study of the wellbore stability of oil and gas wells, which can give us ideas on solving the problems of a broken coal seam. Zhang et al. [11] proposed the coupling system of both flow and solid in rock failure process analysis, in order to study how the size-effect of a horizontal wellbore in a coal seam impacts on wellbore stability by RFPA^{2D}. Chu and Shen et al. [12–14] proposed fracture mechanics and a 3-D discrete element method to analyze the mechanical properties of the coal bed methane wellbore stability and the method to determine the window of drilling fluid density. Since the coal seam was non-continuum, Chen et al. [15] established the discrete element model of the collapse pressure of a coal seam based on non-continuum mechanics considering mud intrusion strata. Zhu et al. [16] proposed the model of coal seam wellbore stability based on the discrete element method. Qu et al. [17] presented the evaluation model of borehole stability in a coal seam by using a stress intensity factor to describe the concentrative degree of crack stress. Zhao and Zhang et al. [18,19] concluded the stress field expression around the broken coal seam wellbore by considering the cleat as the crack and computing the additional stress fields of cracks. But those papers did not distinguish the differences between the cracks and cleats of coals and rocks [20]. Cracks are fractures which can easily separate from each other, and cleats are endogenetic fractures, and this means that they contain fillers in the fracture gap space. Through the researches, the impact of the cleat filler and its connected state on the stress field around the entire wellbore were critical [21]. Through reviewing related literatures and materials, cleats are divided into full filled cleats, filled cleats and unfilled cleats [22]. Therefore, the impact on the near stress field of the cleat filler around the entire wellbore could not be eliminated.

According to the current literature review, it was extremely regular that the previous approaches were calculated by simplification; for example, the cleat system included in the above analytical model was extremely regular, and additionally, they did not distinguish the differences between the cleat system and the crack system, which is quite different from the actual cleat. Thus, it would cause a significant error in the result. Given the irregularity of the fracture distribution in broken coal seams and the impact of the cleat filler, the stress field theory model of the broken coal seam wellbore is established, which considers the influence of the cleat filler. In this paper, the stability model of the broken coal seam wellbore is established, which evaluates the block sliding near the wellbore, by introducing the E.MG-C criterion and the limit equilibrium method to analyze the strength failure and tensile resistance of the cleats. The sliding condition of the triquetrous block and the quadrilateral block are also taken into account, and the accuracy of the analytical model is verified through a numerical method by the Particle Flow Code (PFC) software.

2. Stability Model of Fracture Coal Seam Wall

One of the important features of fractured coal seams is that the coal seams contain a large number of cut fractures. These cleat fractures destroy the integrity of the coal seams, as shown in Figure 1, and cut the coal seams into the quadrilateral blocks, trilateral blocks and a few polygonal blocks. As shown in Figure 1, the quadrilateral block includes the block 1, 7, 8, and the triangular block includes the block 2, 4, 6, 9. As their falling depends on whether their adjacent blocks fall off, the block 3 and 5 are not taken into account. Considering that the axial deformation of the borehole is approximately 0, the stress field around the borehole can be simplified to the plane strain state. In the plane, the wall surrounding rock is affected by non-uniform stress σ_H, σ_h, drilling fluid pressure P_0 and the induced stress caused by the cracks.

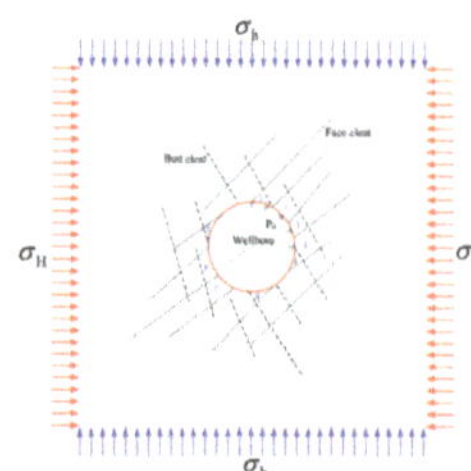

Figure 1. Model of coal wellbore stability.

3. Stress Analysis of the Wellbore Surrounding Rocks in Broken Coal Seams

3.1. Stress Analysis of the Wellbore Surrounding Rocks as a Continuous Body

In order to obtain the stress distribution of broken coal seams, firstly we assume that the coal rock is a continuous body. Then, the stress field around the wall of the well can be obtained under non-uniform, in-situ stress and wellbore pressure P_0:

$$\begin{cases} \sigma'_r = -\sigma_m\left[1-\left(\frac{r}{R_0}\right)^{-2}\right] - \sigma_n\left[1-3\left(\frac{r}{R_0}\right)^{-2}+3\left(\frac{r}{R_0}\right)^{-4}\right]\cos 2\theta - \left(\frac{r}{R_0}\right)^{-2}P_0 - \alpha_{\text{Biot}}P_0 \\ \sigma'_\theta = -\sigma_m\left[1+\left(\frac{r}{R_0}\right)^{-2}\right] + \sigma_n\left[1+3\left(\frac{r}{R_0}\right)^{-4}\right]\cos 2\theta + \left(\frac{r}{R_0}\right)^{-2}P_0 - \alpha_{\text{Biot}}P_0 \\ \tau'_{r\theta} = \sigma_n\left[1+2\left(\frac{r}{R_0}\right)^{-2}-3\left(\frac{r}{R_0}\right)^{-4}\right]\cdot\sin 2\theta \end{cases} \quad , \tag{1}$$

where σ_r, σ_θ, $\tau_{r\theta}$, are radial stress, cyclic stres, and shear stress under polar coordinates, respectively. $2\sigma_m = \sigma_H + \sigma_h$, $2\sigma_n = \sigma_H - \sigma_h$; R_0 is the wellbore radius; r is the length of the calculation point to the wellbore central axis; θ is the wellbore angle; α_{Biot} is Biot coefficient.

Then wellbore rock displacement [23] can be expressed as follows, and the specific derivation process of the formula can be referred to paper 23:

$$\begin{cases} u'_r = \dfrac{1+\mu}{r\cdot E}\dfrac{R_2^2 R_0^2(P_0-(1-2\mu)\sigma_m)+(1-2\mu)P_0 r^2 - R_2^2\sigma_m r^2}{R_2^2-R_0^2} + \\ \qquad\quad \dfrac{2\cos 2\theta}{5E}\left[(12-6\mu)K_1 r^3 - (1+\mu)K_2 r + (6+4\mu)K_3/r + 9(1+\mu)K_4/r^3\right] \\ u'_\theta = \dfrac{6\sin 2\theta}{5E}\left[(3+\mu)K_1 r^3 + (1+\mu)K_2 r - (1-\mu)K_3/r + (1+\mu)K_4/r^3\right] \end{cases} \quad , \tag{2}$$

where E, μ are the elastic modulus and Poisson's ratio of coal seam under plane strain conditions, respectively; u'_r, u'_θ are the radial displacement and the cyclic displacement, respectively; R_2 is the radius of the circular outer boundary set for the solution, $R_2 \gg R_0$; K is a constant coefficient determined by the geometric size of the model. Its specific calculation formula can be written as Equation (3):

$$\begin{bmatrix} 0 & 1 & \frac{2}{R_0^2} & \frac{3}{R_0^4} \\ 3R_0^2 & 1 & \frac{-1}{R_0^2} & \frac{-3}{R_0^4} \\ 0 & 2 & \frac{4}{R_2^2} & \frac{6}{R_2^4} \\ 6R_2^2 & 2 & \frac{-2}{R_2^2} & \frac{-6}{R_2^4} \end{bmatrix}\begin{Bmatrix} K_1 \\ K_2 \\ K_3 \\ K_4 \end{Bmatrix} = \begin{Bmatrix} 0 \\ 0 \\ \sigma_n \\ \sigma_n \end{Bmatrix}, \tag{3}$$

The displacement field under the polar coordinates obtained by Equation (2) can be changed to the displacement field under the orthogonal Cartesian coordinate system by the coordinate transformation method. The transformed displacement component can be expressed as u'_x and u'_y, and the transformation process is shown in Equation (4).

$$\begin{Bmatrix} u'_x \\ u'_y \end{Bmatrix} = \begin{bmatrix} \cos\theta & -\sin\theta \\ \sin\theta & \cos\theta \end{bmatrix}\begin{Bmatrix} u'_r \\ u'_\theta \end{Bmatrix}, \tag{4}$$

3.2. Induced Stress Analysis

According to the research, we find that the internal filling material is not completely filled, and it appears to be intermittent along the orientation of the cleat direction. Therefore, it could be assumed that the filler is composed of multiple rod structures, which is like a spring, as shown in Figure 2. There is no connection between these rods. The strong effect of the rod structure is only connected with the upper side Γ_1 and the lower side Γ_2 on the cleat. When the normal displacement occurs on both sides of the cleat, the filler can be considered as a rod structure. When the cleat surface undergoes a tangent displacement, the filler can be considered as a beam structure.

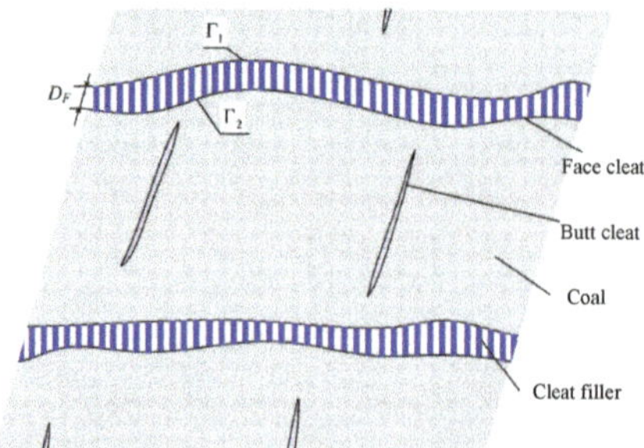

Figure 2. Mechanical model of the cleat filler.

Therefore, under the external loading, the relationship between the filler and the normal load P and the tangent load Q generated on the cleat surface, and the relative displacement on both sides of the cleat surface, can be written as [24]:

$$P = E_F \cdot \frac{\left|u_{y'}^{\Gamma_1} - u_{y'}^{\Gamma_2}\right|}{D_F}, \tag{5}$$

$$Q = \frac{E_F d_b}{4} \cdot \frac{\left|u_{x'}^{\Gamma_1} - u_{x'}^{\Gamma_2}\right|}{D_F^3}, \tag{6}$$

where, E_F is the elastic modulus of the cleat filler; x', y' are directions along the cleat and the vertical cleat, respectively; $u_{x'}$ and $u_{y'}$ are displacements in the x' and y' directions, respectively; d_b is the width of the filler, which can be set as the width of the unit; due to the length of the cleat, along the cleat direction, the normal load P and the tangent load Q will change with different positions, so the cleat load and the cleat distribution load are the function of positions, seen in Figures 3 and 4.

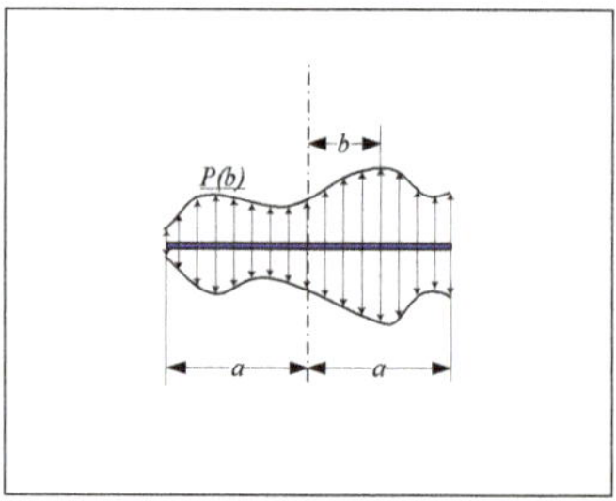

Figure 3. A normal load model with asymmetric distribution for cleats.

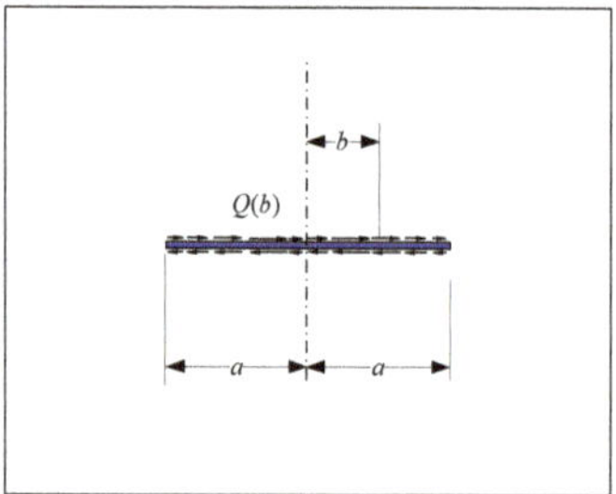

Figure 4. A tangential load model with asymmetric distribution for cleats.

According to the strength factor under a single normal load and a tangent load in the stress field intensity factor manual, and using the superposition principle, the strength factor under an asymmetric distribution load is obtained by integrating [25], seen in Equations (6)and (7):

$$K_I = \sqrt{\frac{1}{\pi a}} \int_{-a}^{a} \sqrt{\frac{a+b}{a-b}} p(b)db,$$ (7)

$$K_{II} = \sqrt{\frac{1}{\pi a}} \int_{-a}^{a} \sqrt{\frac{a+b}{a-b}} Q(b)db,$$ (8)

Then the stress field expression near the crack tip can be written [25] in Equation (9):

$$\begin{cases} \sigma''_{x'} = \frac{K_I}{\sqrt{2\pi\rho}} \cos\frac{\varphi}{2}\left(1 - \sin\frac{\varphi}{2}\sin\frac{3\varphi}{2}\right) - \frac{K_{II}}{\sqrt{2\pi\rho}} \sin\frac{\varphi}{2}\left(2 + \cos\frac{\varphi}{2}\cos\frac{3\varphi}{2}\right) \\ \sigma''_{y'} = \frac{K_I}{\sqrt{2\pi\rho}} \cos\frac{\varphi}{2}\left(1 + \sin\frac{\varphi}{2}\sin\frac{3\varphi}{2}\right) + \frac{K_{II}}{\sqrt{2\pi\rho}} \sin\frac{\varphi}{2}\cos\frac{\varphi}{2}\cos\frac{3\varphi}{2} \\ \tau''_{x'y'} = \frac{K_I}{\sqrt{2\pi\rho}} \sin\frac{\varphi}{2}\cos\frac{\varphi}{2}\cos\frac{3\varphi}{2} + \frac{K_{II}}{\sqrt{2\pi\rho}} \cos\frac{\varphi}{2}\left(1 - \sin\frac{\varphi}{2}\sin\frac{3\varphi}{2}\right) \end{cases},$$ (9)

where, σ''_x, σ''_y and τ''_{xy} are the normal stress and the shearing stress in the additional stress field, respectively. The stress state is shown in Figure 5, where ρ, φ are the radius and angle of the point to the cleat tip, respectively.

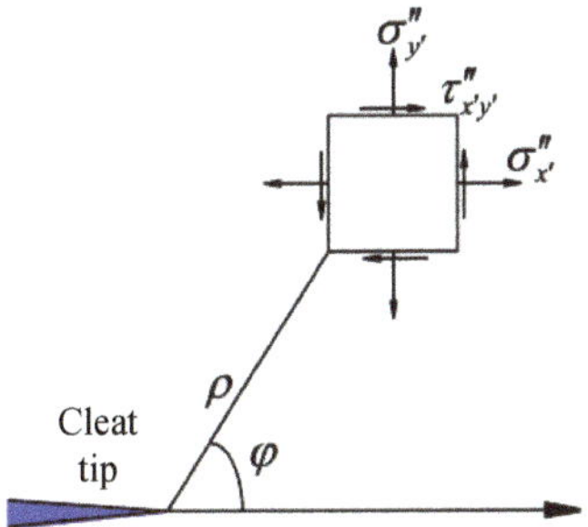

Figure 5. Stress field near the cleat tip in plane problems.

The induced displacement field distribution near the cutting tip can be written as [23]:

$$\begin{cases} u''_x = \frac{K_I}{2G}\sqrt{\frac{\rho}{2\pi}} \cos\frac{\varphi}{2}\left(\kappa - 1 + 2\sin^2\frac{\varphi}{2}\right) + \frac{K_{II}}{2G}\sqrt{\frac{\rho}{2\pi}} \sin\frac{\varphi}{2}\left(\kappa + 1 + 2\cos^2\frac{\varphi}{2}\right) \\ u''_y = \frac{K_I}{2G}\sqrt{\frac{\rho}{2\pi}} \sin\frac{\varphi}{2}\left(\kappa + 1 - 2\cos^2\frac{\varphi}{2}\right) + \frac{K_{II}}{2G}\sqrt{\frac{\rho}{2\pi}} \cos\frac{\varphi}{2}\left(\kappa - 1 - 2\sin^2\frac{\varphi}{2}\right) \end{cases},$$ (10)

where G represents the shear elastic modulus; $\kappa = \frac{3-\mu}{1+\mu}$.

Thus, it can be obtained that the total displacement field of coal seam considering the cleats can be expressed as:

$$\begin{cases} u_x = u'_x + u''_x \\ u_y = u'_y + u''_y \end{cases},$$ (11)

Through the change of coordinates, the relationship between the displacement field and the total displacement field of coal rocks can be obtained:

$$\begin{Bmatrix} u_{x'} \\ u_{y'} \end{Bmatrix} = \begin{bmatrix} \cos\phi & \sin\phi \\ -\sin\phi & \cos\phi \end{bmatrix} \begin{Bmatrix} u_x \\ u_y \end{Bmatrix},$$ (12)

where ϕ is the cutting inclination.

Through the simultaneous Equations (2), (5), (8), (10) and (11), the specific solution can be obtained. In this paper, we propose a program to gain the result of the above formula, because Equations (5)–(8) and (10) keep positive correlation with displacement $u_{x'}$, $u_{y'}$. So, we firstly set the values of $u_{x'}$ and $u_{y'}$, and next through the above formula, we can get other values of $u_{x'}$ and $u_{y'}$. Then taking the result of the previous iteration as the initial value of the continued iteration, we stop iterating until the error between the former and the latter is within the allowable range.

3.3. The Total Stress Field of the Broken Coal Seam Wellbore Surrounding Rocks

The Equations (1) and (9) can be obtained by the stress field distribution of a non-continuous coal seam containing a single shearing. Equations (1) and (9) are in different coordinate systems. For this purpose, the calculation results need to be unified into the same coordinate system for superposition. The coordinate system of the axis X and Y with the σ_H and σ_h directions is the final coordinate system. The stress in Equations (1) and (9) are all converted to this coordinate system, and the stress expression of the non-continuous coal seam is obtained by superposition.

By extension, if n cleats exist in the coal seam, the stress States of σ_x, σ_y and τ_{xy} at any point in the coal seam can be expressed as:

$$
\left\{ \begin{array}{c} \sigma_x \\ \sigma_y \\ \tau_{xy} \end{array} \right\} = \left[\begin{array}{ccc} \cos^2\theta & \sin^2\theta & -\sin 2\theta \\ \sin^2\theta & \cos^2\theta & \sin 2\theta \\ \frac{\sin 2\theta}{2} & -\frac{\sin 2\theta}{2} & \cos 2\theta \end{array} \right] \left\{ \begin{array}{c} \sigma'_r \\ \sigma'_\theta \\ \tau'_{r\theta} \end{array} \right\} + \sum_{i=1}^{n} \left[\begin{array}{ccc} \cos^2\phi_i & \sin^2\phi_i & -\sin 2\phi_i \\ \sin^2\phi_i & \cos^2\phi_i & \sin 2\phi_i \\ \frac{\sin 2\phi_i}{2} & -\frac{\sin 2\phi_i}{2} & \cos 2\phi_i \end{array} \right] \left\{ \begin{array}{c} \sigma''^{i}_{x'} \\ \sigma''^{i}_{y'} \\ \tau''^{i}_{x'y'} \end{array} \right\}, \quad (13)
$$

where $\sigma''^{i}_{x'}$, $\sigma''^{i}_{y'}$, and $\tau''^{i}_{x'y'}$ are the induced stress fields caused by the i cutting, respectively; ϕ_i is the direction of section i.

4. Stable Analysis of Wellbore Coal Rocks

4.1. The Theory of Block Slip and the Determination of Movable Block

The rock in the engineering structure tends to be divided into different shapes by faults, joints, or weak interlayers. Because the strength of the joints and weak interlayers is weaker than the strength of the rock body, the damage of the engineering structure can be attributed to the sliding damage of rocks along the joint surface and the weak interlayer plane, and both the damage and the deformation of rocks are basically ignored.

The block slip theory assumes that the through structure surface is a spatial plane, the mosaic block body is regarded as a convex body, and the various loads of the structure surface are regarded as a spatial vector. Under the conditions of each spatial plane, the geometric method is used to pass simple static force calculation. Then, using the vector operation algorithm, the force of each plane of the mosaic block is summed up by vectors, and the sliding force of the mosaic block instability is obtained. Due to natural cutting, the coal rock is cut into finite rock bodies and infinite rock bodies. The finite rock blocks are divided into movable rock bodies and immovable rock bodies. The key to block theory is to determine the movable blocks.

Through the research, it is found that the parameters that determine the motility of the blocks mainly include the shear plane parameters and the geometric parameters of the near-empty plane. By moving the half-space interface of the constituent block to the coordinate zero point, a series of cones can be formed: the crack cone (CP) is a cone formed by the semi-space of the thermal fracture plane, and the excavation cone (EP) is a cone composed of barely the semi-space of the near-empty plane and the block cone (BP). Combining the block finiteness theorem and the mobility theorem, we can determine the movable block [26] as in Equation (14):

$$
\left\{ \begin{array}{c} CP \neq \varnothing \\ BP = EP \cap CP = \varnothing \end{array} \right. ', \quad (14)
$$

where $\varnothing$ represents an empty set.

4.2. Coal and Rock Cleat Failure Criterion

A large number of rock strength tests show that the strength of rock materials in high stress areas is hyperbolic. Therefore, we use the Extend Mogi-Coulomb (E.MG-C) criterion proposed by A.M. Ai-Ajm [27] as a criterion for the shear strength failure of dry, hot rock cracks. It can be expressed as:

$$\tau_{oct} = a + b\sigma_m + c\sigma_m^2, \tag{15}$$

$$\tau_{oct} = \frac{1}{3}\sqrt{(\sigma_1 - \sigma_2)^2 + (\sigma_2 - \sigma_3)^2 + (\sigma_3 - \sigma_1)^2}, \tag{16}$$

where τ_{oct} is octahedral shear stress; σ_1, σ_2, σ_3 are the first, second and third main stresses respectively; σ_m is the average main stress; a, b are parameters related to the cohesive force and internal friction angle of the rock, respectively, and c is the parameter of the nonlinear nature of the rock strength curve in high stress areas.

The criterion for tensile damage to cracks is shown as:

$$\sigma_t \geq [\sigma_t], \tag{17}$$

where σ_t is the positive tensile stress and $[\sigma_t]$ is an allowable positive stress.

4.3. Block Sliding Direction

The block sliding direction is mainly determined by the structure of the block itself and the force of the block. Assuming that α is the angle between the side of the block and the direction $\vec{k}$, which is from the center of the block to the center of the wellbore, and α is marked as positive in a clockwise direction from the direction $\vec{k}$, α_{bc}^i is the angle between the boundary i and $\vec{k}$, and the sliding angle range determined by the block's own structure can be expressed as:

$$\min\{\alpha_{bc}^i\} \leq \alpha_{sl} \leq \max\{\alpha_{bc}^i\}, \tag{18}$$

where α_{sl} is the sliding angle of the block; $\min\{\alpha_{bc}^i\}$, $\max\{\alpha_{bc}^i\}$ are the minimum and maximum values of the α_{bc}^i composition set when it can slide alone along the boundary i.

The equation for calculating the force on the cleat surface and the weak sandwich plane of the movable block is:

$$\begin{cases} F_{bc}^{ix'} = \sum_{j=1}^{n} \tau_j l_j d \\ F_{bc}^{iy'} = d \sum_{j=1}^{n} \sigma_j l_j d \end{cases}, \tag{19}$$

where $F_{bc}^{ix'}$ and $F_{bc}^{iy'}$ are the combined forces along the x' and y' directions on the boundary I, respectively. x' direction is along the boundary i of the block, and y' direction is perpendicular to the boundary i of the block; n is the number of segments of the boundary i of the block. Since the weak surfaces such as cracks, cleat and weak interlayers, are generally very long, the stress is unevenly distributed along the thin and weak surfaces, and the calculation accuracy of the result by summation after segmentation calculation is higher in solid segments; σ_j, τ_j, l_j are the positive stresses, shear stresses, and lengths of segment j above the boundary i respectively; d is for the thickness of weak surfaces such as cracks, joints and the weak mezzanine.

The equation for calculating the force on the thermal fracture of a movable block can be shown as:

$$\begin{cases} F_{bc}^{ix'} = F_{bc}^{iy'} f \\ F_{bc}^{iy'} = d \sum_{j=1}^{n} \sigma_j l_j d \end{cases} ,$$ (20)

where f is the coefficient of the friction on the surface of the hot fracture.

Combining the above two equations, the combined force on the boundary i of the block is:

$$\overrightarrow{F}_{bc}^{i} = F_{bc}^{ix'} \overrightarrow{i'} + F_{bc}^{iy'} \overrightarrow{j'} ,$$ (21)

Assuming the x direction is along the direction $\overrightarrow{k}$, and the y direction is perpendicular to the direction $\overrightarrow{k}$, the two components of force $\overrightarrow{F}_{bc}^{i}$ are calculated as:

$$\begin{Bmatrix} F_{bc}^{ix} \\ F_{bc}^{iy} \end{Bmatrix} = \begin{bmatrix} \cos \alpha_{bc}^{i} & \sin \alpha_{bc}^{i} \\ -\sin \alpha_{bc}^{i} & \cos \alpha_{bc}^{i} \end{bmatrix} \begin{Bmatrix} F_{bc}^{ix'} \\ F_{bc}^{iy'} \end{Bmatrix} ,$$ (22)

The resultant of forces $\overrightarrow{F}_{bc}$ on the entire block is:

$$\overrightarrow{F}_{bc} = \sum_{i=1}^{N} \overrightarrow{F}_{bc}^{i} ,$$ (23)

where N is the number of boundaries on the block. The size and direction of the joint force $\overrightarrow{F}_{bc}$ on the block can be expressed as:

$$F_{bc} = \sqrt{\left(\sum F_{bc}^{ix}\right)^2 + \left(\sum F_{bc}^{iy}\right)^2} , \tan \gamma_f = \frac{\sum F_{bc}^{iy}}{\sum F_{bc}^{ix}} ,$$ (24)

where F_{bc} is the size of the resultant of forces; γ_f is the angle between the resultant of forces and the direction $\overrightarrow{k}$.

If $\gamma_f \in \left[\min\{\alpha_{bc}^{i}\}, \max\{\alpha_{bc}^{i}\}\right]$, the block slides along the angle γ_f in the direction $\overrightarrow{k}$; when $\gamma_f \notin \left[\min\{\alpha_{bc}^{i}\}, \max\{\alpha_{bc}^{i}\}\right]$, if $\left|\gamma_f - \min\{\alpha_{bc}^{i}\}\right| < \left|\gamma_f - \max\{\alpha_{bc}^{i}\}\right|$, the block slides along the angle $\min\{\alpha_{bc}^{i}\}$ in the direction $\overrightarrow{k}$, or else it slides down the angle $\max\{\alpha_{bc}^{i}\}$ in the direction $\overrightarrow{k}$.

5. Case Analysis

Taking vertical well HG61-4-X in the Qinshui Basin (Shanxi) as an example, it is affected by the fault zone and the tectonic stress around. Therefore, the cleats grow extremely. During the drilling, there is a collapse in the range of 834.90–845.12 m in the coal seam. The numerical model of the coal seam blocks is established based on the slide theoretical model. As shown in Figure 6a, quadrilateral ABCD is the quadrilateral block. As shown in Figure 6b, triangle AED is the triquetrous block. Coal and rock mechanics parameters are shown in Table 1. The basic parameters are listed as follows: the depth is 841 m, the maximum horizontal principal stress $\sigma_H = 17.87$ MPa, the minimum horizontal principal stress $\sigma_h = 16.68$ MPa, the drilling fluid pressure $P_0 = 8.3$ MPa, the formation pore pressure is 8.15 MPa and the borehole diameter $R_0 = 153.1$ mm.

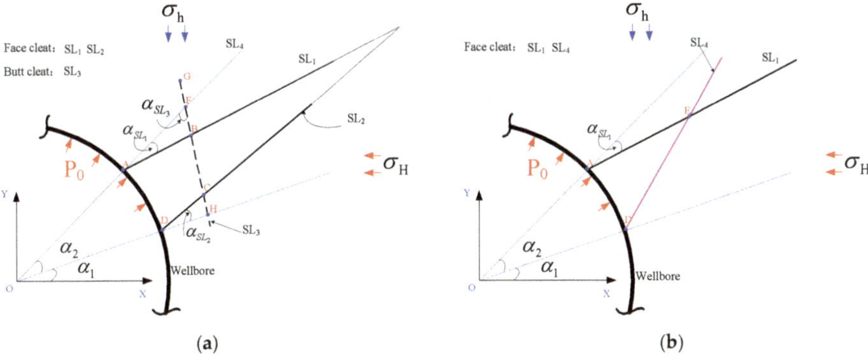

Figure 6. Calculation Model. α_{SL_1}, α_{SL_2}, α_{SL_3} are geometric angles of each cleat respectively; α_1, α_2 are geometric position controlled angles of the block respectively; SL_1, SL_2, SL_4 are face cleats, SL_3 is butt cleat. (**a**) the quadrilateral block, (**b**) the triquetrous block.

Table 1. Mechanic parameters of coal.

Coal and Rock Mechanics Parameters	Value
Elastic modulus E/GPa	10.15
Poisson ratio μ	0.25
Biot coefficient αBiot	0.76
Tensile strength St/MPa	0.56
Elastic modulus of cleat filler Ef/GPa	1.015
Face cleat tensile strength Stm/MPa	0.274
Butt cleat tensile strength Std/MPa	0.426
a	1.613
b	0.864
c	−0.0014

5.1. The Stress Field Comparative Analysis of Coal Seams Containing Multiple Cleats

In order to verify the accuracy of the coal seam stress field containing the multiple cleats analytical solution, the identical mechanical parameters are used, and the finite element model is established. Meanwhile, the principal stress map is obtained, as shown in Figure 7.

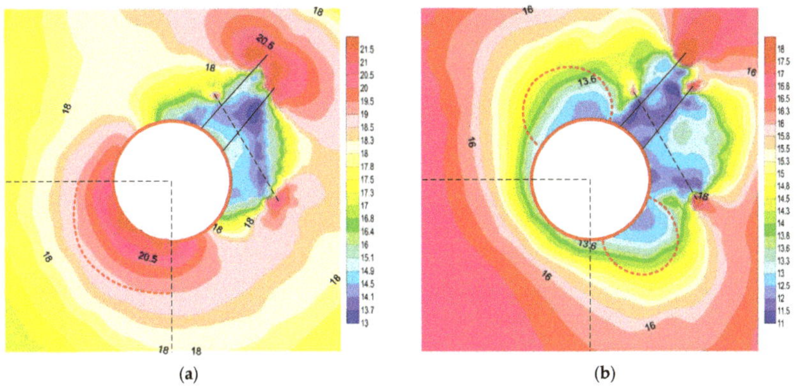

Figure 7. The main stress map of coal bed containing multiple cleats gained by field-emission microscopy (FEM). (**a**) The maximum principal stress map. (**b**) The minimum principal stress map.

As shown in Figure 7, the existence of cleats makes the stress yielded around the coal bed in the wellbore extraordinarily complicated. According to the calculation consequence, the stress value of any point in the coal seam is vitally impacted by the cleats around. The following assumption is made, that is, the stress values are completely different among two points which are closed in the coal seam. To verify the accuracy of the analytical solution, the finite element results in the wellbore should be taken into account. Additionally, after calculation of the consequence at the counterpart position, the curve comparison is shown in Figure 8.

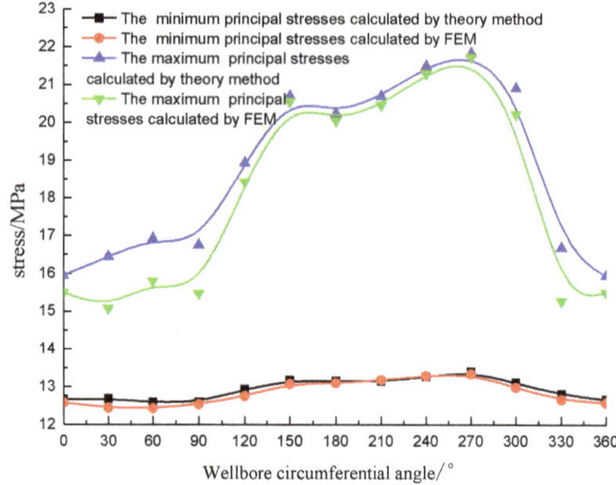

Figure 8. The main stress curves of the wellbore gained by FEM and the theory method.

According to Figure 8, the maximum and minimum principal stresses calculated by the theory method are slightly larger than those calculated by the finite element method, and the relative error of those methods ranges from 0.64% to 9.16%. When the round angle ranges from 120° to 300°, the error is minimum. Based on these analyses, the proposed analytic model of the coal bed stress yield is reliable.

5.2. The Wellborn Stability Analysis of the Coal Seam

η_{SH}, η_{EQ} are defined as block strength coefficient and equilibrium slide coefficient, respectively, in this paper.

$$\eta_{SH} = \frac{\tau_{oct}}{a + b\sigma_m + c\sigma_m^2}, \tag{25}$$

$$\eta_{EQ} = \frac{F_{EQ}}{S_t l}, \tag{26}$$

If $\eta_{SH} \geq 1$, the cleats present strength failure, but on the contrary, the cleats do not present strength failure. If $\eta_{EQ} \geq 1$, the load balance of the block along the slide is broken, which means that the block begins sliding. Otherwise, the load balance of the block along its sliding is maintained. For the quadrilateral block, the only way to slide is by overcoming the strength in both sides along the sliding, and breaking the limit equilibrium in the side which is perpendicular to the sliding direction. So, there are three conditions required to judge the falling of the quadrilateral block. For the triquetrous block, the only way to slide is in overcoming the strength fracture on one side and breaking the limit equilibrium on the other side. Therefore, there are two conditions required to judge the falling of the triquetrous block.

In order to verify the accuracy of the analytical model, the numerical model is established by PFC software. The boundary condition (B.C.) of the numerical model is the same as that of the analytical model. The PFC numerical model is presented in Figure 9, and the red lines represent the face cleats, while the blue lines represent the butt cleats.

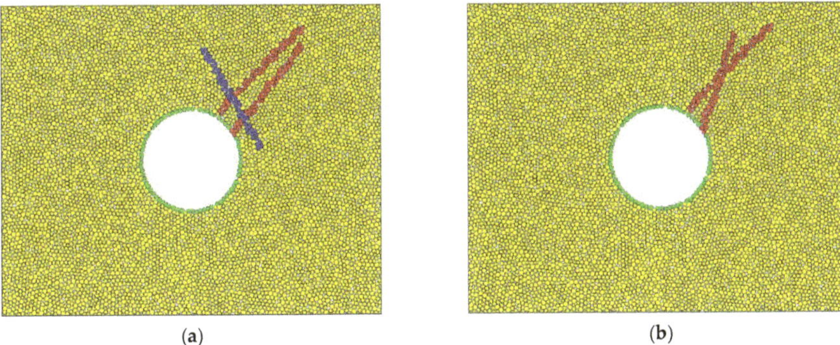

(a)   (b)

Figure 9. Discrete element method (DEM) model by the Particle Flow Code (PFC) software. (a) The quadrilateral block. (b) The triquetrous block.

As seen from Figure 10, the PFC has calculated the sliding result of the block around the wellbore. The calculated result shows that the sliding coefficients of the analytical model prediction are listed as follows: the quadrilateral block, $\eta_{SH_1} = 1.58$, $\eta_{SH_2} = 3.31$, $\eta_{EQ} = 69.79$; the triquetrous block, $\eta_{SH} = 2.49$, $\eta_{EQ} = 1.43$. Since all sliding coefficients are greater than 1, it indicates that the block will fall off. The prediction made by the discrete element method is in accord with the theoretical prediction. To further verify the accuracy of the analytical model, the minimum drilling fluid pressure of maintaining the stability of the wellbore will be calculated spontaneously by the analytical model and the discrete element model. From the results, for the quadrilateral block, the minimum drilling fluid pressure by the analytical mode is 9.4 MPa, and that by the discrete element model is 9.7 MPa and the gap is 3.19%; for the triquetrous block, the minimum drilling fluid pressure by the analytical model is 11.1 MPa, and that by the discrete element is 11.5 MPa, and the gap is 2.68%. Those results indicate that the analytical model and the discrete element model keep highly identical.

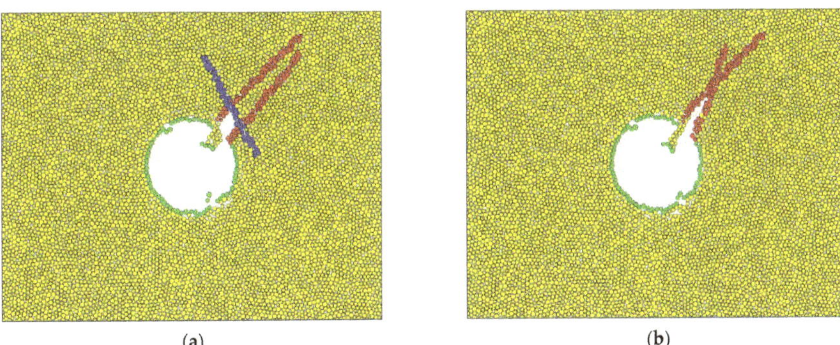

(a)   (b)

Figure 10. Slip map of the coal block near the wellbore by numerical simulation. (a) The quadrilateral block. (b) The triquetrous block.

5.3. The Effect Factor Analysis

The stability of the coal seam wellbore is impacted greatly by the geometry and the position of the block. Therefore, this paper focuses on discussing how the face cleat spacing, the butt cleat spacing, the cleat length, the cleat inclination angle and the block geometric position have an effect upon the strength sliding coefficient and the equilibrium slip coefficient of the quadrilateral block and the triquetrous block. The calculation parameters and the analysis parameters can be referred to Figure 6.

5.3.1. Face Cleat Spacing

In order to analyze the influence of the face cleat spacing for block sliding coefficient, the face cleat spacing between SL_1 and SL_2 is modified, keeping the other parameters invariant. The face cleat spacing ranges from 2.09 cm to 3.19 cm.

According to Figure 11, the addition of the face cleat distance will cause the decrease of the strength coefficient and equilibrium slip coefficient of the quadrilateral block, and the decrease of η_{SH2} is greater than η_{SH1}. When the distance is above 3.13 cm, the quadrilateral block stops falling off. On the contrary, the addition of the face cleat distance will cause the increase of the strength coefficient and the equilibrium slip coefficient of the triangular block, and the increase of η_{SH2} is greater than that of η_{SH1}. When the distance is above 2.87 cm, the triangular block starts to drop. The increase of the face cleat distance decreases the interference among the cleats. Additionally, it reduces the stresses of the lines AB and CD on the face cleats, and it causes the reduction of the strength slip coefficient. The difference of the direction and the length of the face cleats SL_1 and SL_2 are both the reason why the ranges of the discount are different between η_{SH2} and η_{SH1}. The increase of the face cleat distance causes the addition of the resistance to the triquetrous block sliding, and it reduces the un-equilibrium force in blocks, so the equilibrium slip coefficient decreases. For the triquetrous block, the shape becomes blunt and wide from a cuspidal, and the face cleat distance becomes narrow. Meanwhile, the angle of cleat SL_4 decreases. Based on Equations (4)–(6), with the increase of shearing stress and tensile stress on the cleat SL_4, the stresses on cleats AE, DE, the strength coefficient and the equilibrium slip coefficient, all increase.

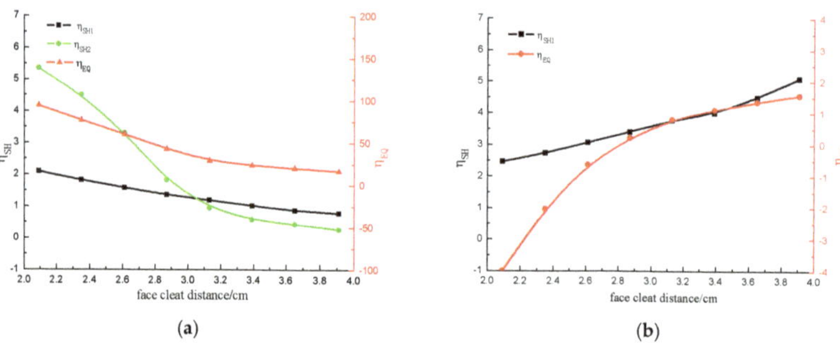

Figure 11. Influence of the face cleat distance on the slip coefficient distribution. (**a**) the quadrilateral block. (**b**) the triquetrous block.

5.3.2. Butt Cleat Distance

In order to analyze how the butt cleat distance makes an impact on the block slip coefficient, the length of line AF is modified, keeping other factors invariant. The length of the line AF ranges from 4 cm to 14 cm.

According to Figure 12, with the increase of the butt cleat distance, the strength slip coefficient of the quadrilateral block increases, and the equilibrium slip coefficient of the quadrilateral block climbs up and then declines. But for the triquetrous block, the strength slip coefficient and the equilibrium slip coefficient decline. When the butt cleat distance is longer than 5 cm, the triquetrous block stops falling off. The reason is that under the local coordinate system of cleat SL_3, the dimensionless radius among the midpoints of cleats AB and CD increases as the addition of the butt cleat distance. In addition, this causes the increase of the stress on cleats AB and CD, and at the same time, it affects the increase of the strength slip coefficient. It is hard to conclude the change law of the equilibrium slip coefficient, because the change of the butt cleat distance causes the stress changes on all the sides of the quadrilateral block. Therefore, the calculation is required to define these values. Because of the addition of the butt cleat distance, the body areas of the quadrilateral block and the triquetrous block increase, as well as the angle of cleat SL_4. By Equations (4) and (5), we can know that it causes the decline of the shearing stresses and the tensile stresses on cleat SL_4, as well as stresses on cleats AE and DE. As a result, the strength coefficient and equilibrium slip coefficient decline.

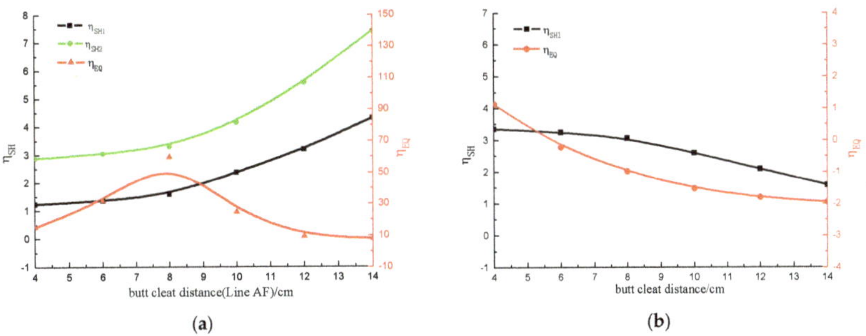

Figure 12. Influence of the butt cleat distance on slip coefficient distribution. (**a**) The quadrilateral block. (**b**) The triquetrous block.

5.3.3. Cleat Length

Keeping other parameters invariant, we change cleat SL_1 length and analyze the effect of the cleat length on the sliding coefficient of the block. SL_1 ranges from 15 cm to 24 cm.

From Figure 13, it can be seen that the increase in the length of the cleat SL_1 reduces the strength and equilibrium sliding coefficient. When the length of the cleat SL_1 is greater than 21 cm, the quadrilateral block stops falling off, and the triquetrous block never falls off. The increase in the length of the cleat reduces the dimensionless radius of the midpoints of the cleat, thereby reducing the stress of the midpoints of the cleat, so the strength slip coefficient decreases. However, the internal pressure of the well wall has not changed. From the limit equilibrium equation, it is known that the decrease of the block slide force is larger than that of the resistant force, so the uneven force of the block decreases and then the equilibrium slip coefficient decreases.

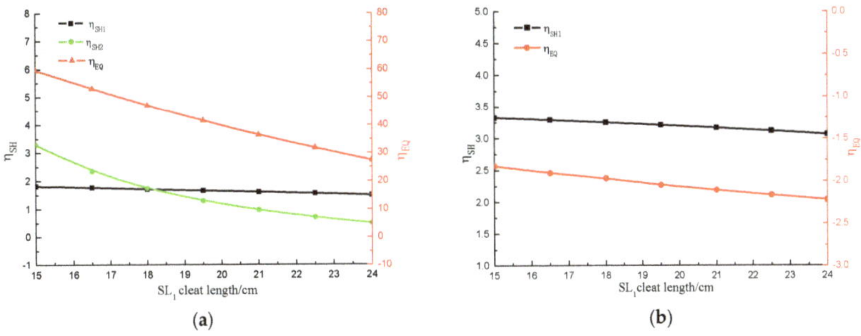

Figure 13. Influence of the cleat length on the slip coefficient distribution. (**a**) The quadrilateral block. (**b**) The triquetrous block.

5.3.4. Cleat Inclination

Keeping other parameters invariant, we change the inclination α_{SL_2} of the cleat SL_2 and analyze the effect of the cleat inclination on the sliding coefficient of the block. Inclination α_{SL_2} ranges from 15° to 21°.

From Figure 14, it can be seen that with the increase of the inclination of the cleat, the strength slip coefficient of the quadrilateral block increases, the equilibrium slip coefficient decreases, and the strength and equilibrium slip coefficient of the triangular block increases. When $\alpha_{SL_2} \geq 20°$, the triquetrous block changes start falling off. As α_{SL_2} increases, from Equations (4)–(6), it can be seen that the shear stress and the tensile stress on the cleat SL_2 increase, and this results in an increase in the stress on the cleat AB and CD, and an increase in the strength slip coefficient of the quadrilateral block. Similarly, the strength slip coefficient of the triquetrous block increases as α_{SL_2} increases. The increase of α_{SL_2} shortens the length of the cleat BC, and reduces the block slide force, so the equilibrium slip coefficient of the quadrilateral block is reduced; and for the triquetrous block, the effect of the α_{SL_2} increase on the strength and equilibrium slip coefficient is the same as the increase in the distance between face cleats (see Section 5.3.1).

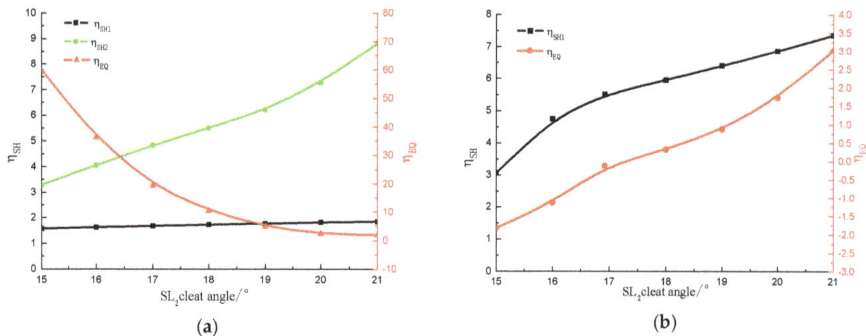

Figure 14. Influence of the cleat angle on the slip coefficient distribution. (**a**) The quadrilateral block. (**b**) The triquetrous block.

5.3.5. Block Geometric Position

Keeping other parameters invariant, we change α_1 and analyze the effect of the block geometry position on the strength and equilibrium slip coefficient. α_1 ranges from 0° to 180°.

From Figure 15a, we can see that with the increase of α_1, the strength slip coefficient and the equilibrium slip coefficient of the quadrilateral block are approximately sinusoidal changes. When α_1 changes from 30° to 120°, the quadrilateral block starts falling off. From Figure 15b, we can also observe that with the increase of α_1, the equilibrium slip coefficient of the triquetrous block is approximately sinusoidal, but the strength slip coefficient is undulating. When α_1 changes from −20° to 10°, and from 60° to 120°, the triquetrous block starts falling off. The change of α_1 causes the change of the cleat angle around the block, which causes the change in the induced stress field and at last causes a fluctuation in the strength and equilibrium slip coefficient.

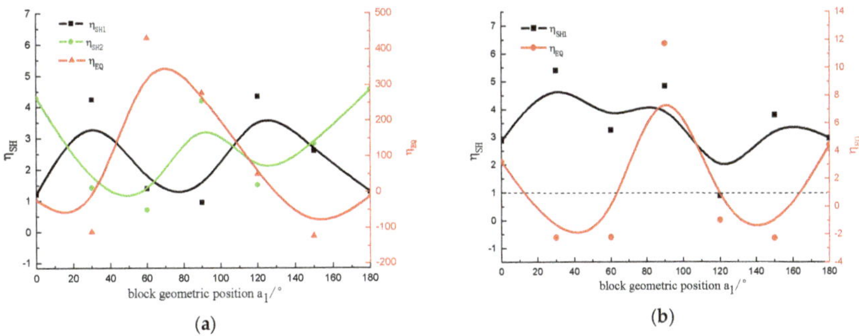

Figure 15. Influence of the block geometric position on the slip coefficient distribution. (**a**) The quadrilateral block. (**b**) The triquetrous block.

6. Discussion

This paper presents a novel stress field analytical model of the wellbore coal rock in the broken coal seam. This analytical model considers the influence of the cleat filler, and by simplifying the cleat filler as the multiple rod structures (see Figure 2), we use the mechanics of materials and fracture mechanics to establish the induced stress field and the displacement field.

According to the limit equilibrium theory and the E.MG-C criterion, we further establish the stable analysis of the wellbore coal rock. The steps are: Firstly, we determine the motility of the block mainly including the shear plane parameters and the geometric parameters of the near-empty plane; secondly, we gain coal and rock cleat failure criterion; and lastly, we gain the block sliding direction. Taking vertical well HG61-4-X in the Qinshui Basin as an example, this paper uses a numerical method (see Figures 7 and 9) to judge the validity of the analytical model.

This paper also analyzes the effect factors such as the face cleat spacing, the butt cleat spacing, the cleat length, the cleat inclination angle and the block geometric position. The results show that a quadrilateral block slides off easier than a triangular block under the same boundary condition; the bigger are the cleat spacing and cleat length, the lower is the risk at which blocks slide off, and the increasing cleat angle could cause blocks to slide off easily. Under the same boundary condition, it is closely related to the well round angle at which blocks slide off.

In addition, we merely consider the drilling fluid fluctuation in this paper. Thus, an improved and more general approach should be proposed in the future work to completely resolve the issue.

7. Conclusions

This paper establishes a stress field analytical model of the wellbore coal rock by considering the irregularity of the cleat distribution and the influence of the cleat filler. According to the limit equilibrium theory and the E.MG-C criterion, the wellbore stability model of the coal seam is established by determining, firstly, the motility of the block, secondly, the cleat failure criterion and thirdly, the block sliding direction. In order to judge the validity of the analytical model, this paper uses a numerical method to make a comparative analysis. At the same time, cleat affecting factors are studied in the paper.

Author Contributions: Conceptualization, X.Z.; methodology, X.Z.; software, Y.M.; validation, X.Z., J.W., and Y.M.; formal analysis, Y.M.; investigation, X.Z., J.W., and Y.M.; resources, Y.M.; data curation, J.W. and Y.M.; writing—original draft preparation, X.Z., J.W., and Y.M.; writing—review and editing, X.Z., J.W., and Y.M.; visualization, Y.M. and X.Z.; supervision, X.Z.; project administration, Y.M.; funding acquisition, X.Z. All authors have read and agreed to the published version of the manuscript.

Funding: This research is funded by the Natural Foundation of Shandong Province, grant number ZR2018BEE012 (Xinbo Zhao). Yue Mei also acknowledges the support from the Fundamental Research Funds for the CentralUniversities (Grant No. DUT19RC(3)017).

Conflicts of Interest: The authors declare no conflict of interest.

References

1. Qin, Y.; Wu, J.G.; Shen, J.; Yang, Z.B.; Shen, Y.L.; Zhang, B. Frontier research of geological technology for coal measure gas joint-mining. *J. China Coal Soc.* **2018**, *43*, 1504–1516.
2. Meng, S.P.; Liu, C.L.; Ji, Y.M. Geological conditions of coalbed methane and shale gas exploitation and their comparison analysis. *J. China Coal Soc.* **2013**, *38*, 728–737.
3. Li, X.F.; Pu, Y.C.; Sun, C.Y.; Ren, W.N.; Li, Y.Y.; Zhang, Y.Q.; Li, J.; Zang, J.L.; Hu, A.M.; Wen, S.M.; et al. Recognition of absorption/desorption theory in coalbed methane reservoir and shale gas reservoir. *Acta Pet. Sin.* **2014**, *35*, 1113–1129.
4. Ai, C.; Hu, C.; Zhang, Y.; Yu, L.; Li, Y.; Wang, F. Wellbore Stability Estimation Model of Horizontal Well in Cleat-Featured Coal Seam. In Proceedings of the SPE/EAGE European Unconventional Resources Conference and Exhibition, Vienna, Austria, 25–27 February 2014.
5. Liang, D.C.; Po, X.L.; Xu, X.H. Particularity of coal collapse and drilling fluid countermeasure. *J. Southwest Pet. Inst.* **2002**, *24*, 28–31.
6. Jianguo, Z.; Yuanfang, C.; Mingbo, S. Experimental study on sloughing mechanism of coal beds in Tarim Oilfield. *Drill. Pet. Tech.* **2000**, *28*, 21–23.
7. AI-Ajmi, A.M.; Sultan Qaboos, U.; Zimmerman, R.W. Stability Analysis of Deviated Boreholes Using the Mogi-Coulomb Failure Criterion, With Applications to Some Oil and Gas Reservoirs. In Proceedings of the IADC/SPE Asia Pacific Drilling Technology Conference 2006—Meeting the Value Challenge: Performance, Deliverability and Cost, Bangkok, Thailand, 13–15 November 2006.
8. Gallant, C.; Zhang, J.; Wolfe, C.A.; Freeman, J.; Al-Bazali, T.M.; Reese, M. Wellbore Stability Considerations for Drilling High-Angle Wells through Finely Laminated Shale: A Case Study from Terra Nova. In Proceedings of the SPE Annual Technical Conference and Exhibition, Anaheim, CA, USA, 11–14 November 2007.
9. Jiang, Z.; Zhang, J.; Meng, Y.; Li, G. Analysis on the methods used to evaluate the wellbore stability of gas drilling. *Nat. Gas Ind.* **2007**, *7*, 68–70.
10. Chunhua, G. Simulation of In-Situ Stress near Wellbore and Research on Wellbore Stability-Take Xujiahe Formation Gas Reservoir in Western Sichuan Depression as an Example. Ph.D. Thesis, Chengdu University of Technology, Chengdu, China, 2011.
11. Zhang, Z.; Tang, C.A.; Li, L.C.; Ma, T.H. Numerical investigation on stability of wellbore during coal bed gas mining process. *China Min. Mag.* **2006**, *15*, 55–58.
12. Qu, P.; Shen, R.; Fu, L.; Wang, Z. Time delay effect due to pore pressure changes and existence of cleats on borehole stability in coal seam. *Int. J. Coal Geol.* **2011**, *85*, 212–218. [CrossRef]
13. Ping, Q.; Ruichen, S. Mechanism of wellbore stability and determination of drilling fluid density window in coalbed methane drilling. *Nat. Gas Ind.* **2010**, *30*, 64–68.
14. Qu, P.; Shen, R.C.; Fu, L.; Zhang, X.L.; Yang, H.L. Application of the 3D discrete element method in the wellbore stability of coal-bed horizontal wells. *Acta Pet. Sin.* **2011**, *32*, 153–157.
15. Mian, C.; Haifeng, Z.; Yan, J.; Yunhong, D.; Yonghui, W. A discontinuous medium mechanical model for the sidewall stability prediction of coal beds. *Acta Pet. Sin.* **2003**, *34*, 145–151.
16. Zhu, X.; Liu, W.; Jiang, J. Research regarding coal-bed wellbore stability based on a discrete element model. *Pet. Sci.* **2014**, *11*, 526–531. [CrossRef]
17. Ping, Q.; Ruichen, S.; Henglin, Y. Evaluation model of wellbore stability in coal seam. *Acta Pet. Sin.* **2006**, *38*, 835–842.
18. Zhang, L.; Yan, X.; Yang, X.; Zhao, X. An analytic model of wellbore stability for coal drilling base on limit equilibrium method. *J. China Coal Soc.* **2016**, *41*, 909–916.
19. Zhang, L.; Yan, X.; Yang, X.; Zhao, X. An analytical model of coal wellbore stability based on block limit equilibrium considering irregular distribution of cleats. *Int. J. Coal Geol.* **2015**, *152*, 147–158. [CrossRef]
20. Jiancheng, J.; Hong, Z.; Qian, J.; Yan, W.; Miaofeng, Z.; Chen, C. Status and prospect: Study on the cleat system in coal reservoir. *Nat. Gas Geosci.* **2015**, *26*, 1621–1628.
21. Close, J. Natural fracture in coal. *AAPG Stud. Geol.* **1993**, *38*, 119–132.
22. Shengli, Z. Coal cleat and its significance in the coalbed methane exploration and development. *Coal Geol. Explor.* **1995**, *23*, 27–31.

23. Xinbo, Z.; Xiujuan, Y.; Xiangyang, L.; Lisong, Z.; Xiangzhen, Y. Integrity analysis of high temperature and high pressure wellbores withthermo-structural coupling effects. *Chin. J. Eng.* **2016**, *261*, 13–20.
24. Hongwen, L. *Mechanics of Materials*; Higher Education Press: Beijing, China, 2017.
25. Shiyu, L.; Taiming, H.; Xiangchu, Y. *Introduction of Rock Fracture Mechanics*; University of Science and Technology of China Press: Hefei, China, 2010.
26. Bing, H.; Hui, S.; Xianda, F.; Yu, L. Analysis and research on the whole stability of rock slope based on key block method. *Eng. Technol. Appl.* **2018**, *24*, 171–173.
27. Mingming, Z.; Lixi, L.; Xiangjun, L. Impact analysis of different rock shear failure criteria towellbore collapse pressure. *Chin. J. Rock Mech. Eng.* **2017**, *36*, 3485–3491.

 © 2020 by the authors. Licensee MDPI, Basel, Switzerland. This article is an open access article distributed under the terms and conditions of the Creative Commons Attribution (CC BY) license (http://creativecommons.org/licenses/by/4.0/).

Article

Energy Concepts and Critical Plane for Fatigue Assessment of Ti-6Al-4V Notched Specimens

Camilla Ronchei *, Andrea Carpinteri and Sabrina Vantadori

Department of Engineering & Architecture, University of Parma, Parco Area delle Scienze 181/A, 43124 Parma, Italy; andrea.carpinteri@unipr.it (A.C.); sabrina.vantadori@unipr.it (S.V.)
* Correspondence: camilla.ronchei@unipr.it; Tel.: +39-0521-905923

Received: 30 April 2019; Accepted: 22 May 2019; Published: 27 May 2019

Abstract: In the present paper, the fatigue life assessment of notched structural components is performed by applying a critical plane-based multiaxial fatigue criterion. Such a criterion is formulated by using the control volume concept related to the strain energy density criterion. The verification point is assumed to be at a given distance from the notch tip. Such a distance is taken as a function of the control volume radii around the notch tip under both Mode I and Mode III loading. The accuracy of the present criterion is evaluated through experimental data available in the literature, concerning titanium alloy notched specimens under uniaxial and multiaxial fatigue loading.

Keywords: control volume concept; critical plane approach; fatigue life assessment; severely notched specimens; strain energy density

1. Introduction

From the pioneering work by Jasper at the beginning of the 1920s [1], energy-based criteria have been widely used for estimating multiaxial fatigue lifetime of engineering components subjected to time-varying loading [2–10]. The fundamental idea on which such criteria are based is the assumption that energy (calculated in different ways) is always proportional to fatigue damage.

Among the different energy-based criteria available in the literature [2], those proposed by Garud [3] and Ellyin et al. [4–6] are very interesting. The multiaxial fatigue assessment has to be performed through the cyclic plastic deformation according to Garud, whereas Ellyin et al. argued that both cyclic plastic energy and elastic energy have to be properly taken into account in the fatigue lifetime estimation.

It is worth noting that the fatigue life of notched structural components subjected to cyclic loading can be evaluated by means of energy-based concepts. In particular, the concept of strain energy density (SED) has originally been implemented in different criteria available in the literature to predict the fatigue behavior of notched components under uniaxial tensile loading [7,8]. Subsequently, SED-based criteria have been formulated for multiaxial loading [9,10].

The main drawback of the above criteria is that the fatigue behavior is assumed to depend only on the stresses at the notch tip. Therefore, any SED-based criterion cannot be applied at the tip of sharp notches since both the stress state and SED tend toward infinite.

In order to overcome the above problem, Lazzarin and Zambardi suggested considering a small but finite volume of material close to the notch tip (that is, the point of stress singularity), over which the SED has a finite value [11]. More precisely, the fatigue damage parameter for blunt and sharp notched structural components under tensile loading (Mode I) is the mean value of the SED, related to a control volume around the notch tip [11,12]. The radius of the above volume depends on the unnotched specimen fatigue limit, the notch stress intensity factor (NSIF) range and the elastic Poisson's ratio. The Lazzarin and Zambardi criterion has also been extended to notched structural components subjected to multiaxial loading [13–18] as well as to welded joints [19–21].

Recently, an attempt to perform the fatigue lifetime assessment of notched specimens through energy concepts has been made by Carpinteri and co-workers [22–24], relatively to Ti–6Al–4V titanium alloy specimens under uniaxial and multiaxial loading (biaxiality ratio 0.6 and 2.0), where the control volume concept has been implemented in the original formulation of the critical plane-based multiaxial fatigue criterion [25,26].

In the present paper, the above criterion is proposed to be applied together with the control volume concept, and fatigue assessment is performed in a verification point at a distance related to energy concepts.

It is validated by means of experimental data related to V-notched specimens made of titanium grade 5 alloy, subjected to mixed mode loading [27]. Such material and other titanium alloys have attracted significant interest being extensively used in leading industries, due to their low density and high specific strength at elevated temperature (aeronautics, nuclear energy) and their compatibility with human tissues (applications in the biomedical field) [28–32]. In the latter field, such materials can be used, for example, in the form of β-type titanium porous structures [30,31], and functionally graded Ti-6Al-4V alloy interconnected mesh structures [32].

In Section 2, the theoretical framework of the strain-based criterion by Carpinteri et al. is outlined, also implementing the concept of control volume [22–24]. Then, the validation of such a criterion by means of experimental data for combined tension and torsion cyclic loading on V-notched specimens is shown in Section 3, and finally, the conclusions are drawn in Section 4.

2. Theoretical Framework of Strain-Based Multiaxial Fatigue Criterion

The strain-based multiaxial fatigue criterion by Carpinteri et al. [25,26] is related to the critical plane approach and consists of three steps detailed in the following sub-sections: Step I, where the verification point position (point P) is analytically defined by means of the control volume concept, Step II, where the critical plane orientation is theoretically determined, Step III, where the fatigue life assessment is performed in such a plane at point P. The criterion is suitable to be applied to ductile materials under low cycle fatigue loading.

2.1. Step I: Verification Point Position

The fatigue lifetime is computed at point P which is on the notch bisector at a certain distance, r, from the notch surface [25,26] (Figure 1). The expression of the above distance is obtained by means of a best fitting procedure (details are provided in Reference [24]) and is given by [22]:

$$r = -(0.221)^{\lambda - 1.484} \cdot R_m + 11.3 R_m \tag{1}$$

where λ is the ratio between the amplitude of the remote shear stress and the amplitude of the remote normal stress (named biaxiality ratio) and R_m is computed as the mean value of R_1 (control volume radius under Mode I loading) and R_3 (control volume radius under Mode III loading). The last ones are calculated by the SED criterion, and are functions of notch stress intensity factor (NSIF) ranges (ΔK_{1A}, ΔK_{3A}), high-cycle fatigue strengths of smooth specimens ($\Delta \sigma_{1A}$, $\Delta \tau_{3A}$) and the notch geometry [33]. More precisely, the above radii are computed according to the following equations [33]:

$$R_1 = \left(\sqrt{2e_1} \cdot \frac{\Delta K_{1A}}{\Delta \sigma_{1A}} \right)^{\frac{1}{1 - \lambda_1}} \tag{2a}$$

$$R_3 = \left(\sqrt{\frac{e_3}{1 + v_e}} \cdot \frac{\Delta K_{3A}}{\Delta \tau_{3A}} \right)^{\frac{1}{1 - \lambda_3}} \tag{2b}$$

where e_1 and e_3 are two parameters depending on the V-notch geometry, v_e is the elastic Poisson ratio, and λ_1 and λ_3 are the eigenvalues for Mode I and Mode III, respectively, calculated by means of finite element analysis, as is discussed in Reference [33].

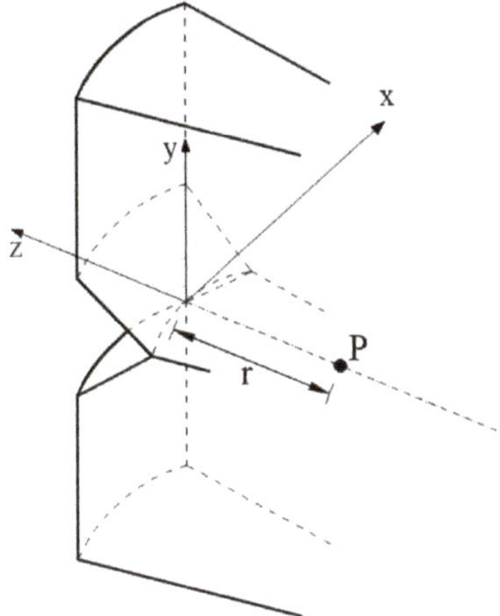

Figure 1. Verification point position according to the control volume concept for V-notch.

A finite element model is employed in order to numerically compute the strain state at verification point P [24]. In particular, linear transient dynamic analyses are performed on notched specimens through the Commercial Package Straus7® [34], by adopting both 6- and 8-node tridimensional finite elements. Only one half of each specimen is modeled, taking advantage of the geometric symmetry. Moreover, the adopted discretization is shown in Figure 2, where the finite element mesh is that adopted after a convergence analysis, being the minimum finite element size equal to about 0.25 times the value of the notch root radius.

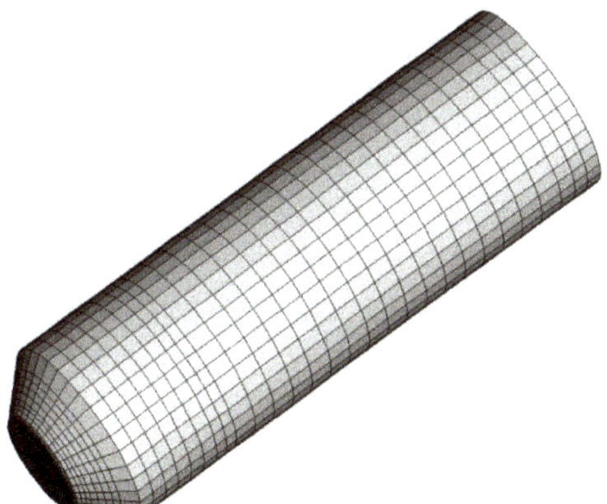

Figure 2. Discretization adopted for the finite element model.

2.2. Step II: Critical Plane Orientation

Let us consider the strain state at point P and a generic time instant t of the fatigue loading history, the principal strain ε_1, ε_2 and ε_3 (with $\varepsilon_1 \geq \varepsilon_2 \geq \varepsilon_3$) and the corresponding directions 1, 2 and 3 (identified by means of the principal Euler angles ϕ, θ and ψ) can be determined. Since the principal directions are usually time-varying under fatigue loading, Carpinteri et al. proposed to compute the averaged directions $\hat{1}$, $\hat{2}$ and $\hat{3}$ on the basis of the instantaneous ones, through the averaged values of the principal Euler angles [35].

Then, the critical plane orientation is regarded to depend on such averaged directions. In more detail, the normal vector w to the critical plane is assumed to be linked to the $\hat{1}$-direction through an off-angle δ, given by the following empirical expression [25,26]:

$$\delta = \frac{3}{2}\left[1 - \left(\frac{1}{2\left(1 + v_{eff}\right)} \frac{\gamma_a}{\varepsilon_a} \right)^2 \right] 45° \tag{3}$$

being v_{eff} the effective Poisson ratio (that is, function of both elastic, v_e, and plastic, v_p, Poisson's ratio), and ε_a and γ_a the strain amplitudes in the well-known tensile and torsional Manson–Coffin equations, respectively. Note that the above rotation (Equation (2)) has to be performed from $\hat{1}$ to $\hat{3}$ in the principal plane $\hat{1}\hat{3}$.

2.3. Step III: Fatigue Life Assessment

The fatigue life assessment is performed through the following expression, where the left-hand term corresponds to an equivalent strain whose amplitude is a function of the amplitudes $\eta_{N,\,a}$ and $\eta_{C,\,a}$ of both the normal and the tangential displacement vectors [25,26]:

$$\varepsilon_{eq,a} = \sqrt{\left(\eta_{N,a}\right)^2 + \left(2\left(1 + v_{eff}\right) \cdot \frac{\varepsilon_a}{\gamma_a}\right)^2 \left(\eta_{C,a}\right)^2} \tag{4}$$

Note that all terms in Equation (3) depend on the number of loading cycles to failure. Moreover, the values of $\eta_{N,\,a}$ and $\eta_{C,\,a}$ are obtained from an analytical procedure by taking into account both the strain tensor at point P and the critical plane orientation. Details can be found in Reference [24].

The fatigue life (i.e. the theoretical number of loading cycles to failure, N_f) is iteratively computed by equaling Equation (4) [24] with the Manson–Coffin normal strain amplitude ε_a.

3. Criterion Validation

In order to check the accuracy of the criterion presented in Section 2, some experimental data are selected from the technical literature [27,33]. Such data are related to uniaxial and multiaxial fatigue tests (with nominal loading ratio equal to −1) carried out on circumferentially V-notched cylindrical specimens characterized by (Figure 3):

- V-notch with a depth of 6 mm;
- Opening angle of 90°;
- Notch root radius of 0.1 mm.

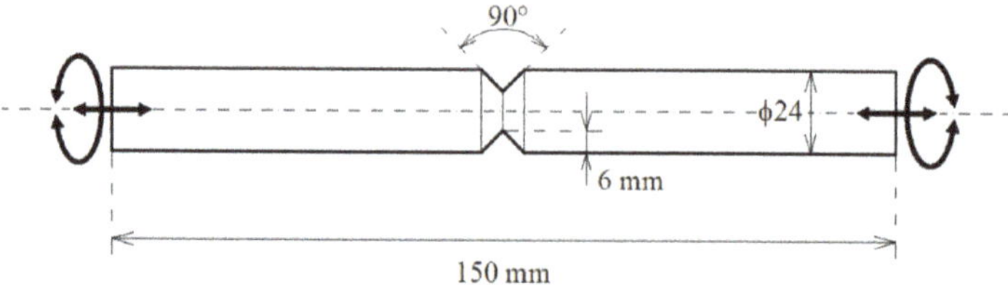

Figure 3. Geometrical sizes of the titanium alloy V-notched specimens subjected to tension and/or torsion fatigue loading.

The above specimens were made of grade 5 titanium alloy (Ti-6Al-4V), commonly used in aerospace and naval applications. The above titanium alloy is characterized by very good static and fatigue properties (Table 1) with a high strength-to-mass ratio.

Table 1. Static and fatigue properties of material examined [27,33].

E [GPa]	ν_e [-]	σ_u [MPa]	σ_y [MPa]	ΔK_{1A} [MPa·mm$^{0.445}$]	ΔK_{3A} [MPa·mm$^{0.333}$]	$\Delta\sigma_{1A}$ [MPa]	$\Delta\tau_{3A}$ [MPa]
110.0	0.3	978.0	894.0	452.0	1216.0	950.0	776.0

The Manson–Coffin parameters of both tensile and torsional equations are reported in Reference [22].

By taking full advantage of ν_e, ΔK_{1A}, ΔK_{3A}, $\Delta\sigma_{1A}$ and $\Delta\tau_{3A}$, the control volume radii R_1 and R_3 are equal to 0.051 mm and 0.837 mm, respectively.

Before being fatigue tested, the specimens have been polished in order to remove surface scratches and marks due to machine tools. Fatigue tests have been performed by means of an MTS 809 servo-hydraulic axial-torsional testing system with a 100 kN axial cell and a 1100 Nm torsion cell. Moreover, all tests have been carried out under load control, with a frequency between 10 and 15 Hz. Details of the loading conditions related to the experimental fatigue tests being examined are reported in Reference [27,33]. In particular, we consider four different loading conditions characterized by experimental fatigue life, $N_{f,\exp}$, between 10^3 and $6\cdot10^5$ loading cycles, and more precisely:

1. Pure tension fatigue loading;
2. Pure torsion fatigue loading;
3. Combined in-phase ($\Phi = 0°$) tension and torsion fatigue loading;
4. Combined out-of-phase ($\Phi = 90°$) tension and torsion fatigue loading.

The biaxiality ratio λ related to multiaxial loading conditions is equal to 2.

According to the above loading conditions, the value of the distance r (Equation (1)) turns out to be:

(a) $r = 1.9 \cdot R_m$ for pure tension fatigue loading ($\lambda = 0$);
(b) $r = 11.3 \cdot R_m$ for pure torsion fatigue loading ($\lambda = \infty$);
(c) $r = 10.8 \cdot R_m$ for combined tension and torsion fatigue loading ($\lambda = 2.0$).

Figure 4 shows experimental fatigue life, $N_{f,\exp}$, plotted against the theoretical one, N_f. In particular, 79% of results is conservative and 63% is included into 3× band. Moreover, we can remark that better estimations are obtained by considering only the in-phase multiaxial fatigue data (Figure 4b) since all the results fall within 3× band, whereas almost all the results related to out-of-phase data are outside the above band (Figure 4c).

In any case, when estimations do not fall within the reference bands, the errors made by the present criterion are, in general, on the conservative side. This strongly supports the idea that the Carpinteri et al. criterion, applied together with the control volume concept, can be used successfully to assess notched components in situations of practical interest, always allowing an adequate margin of safety to be reached.

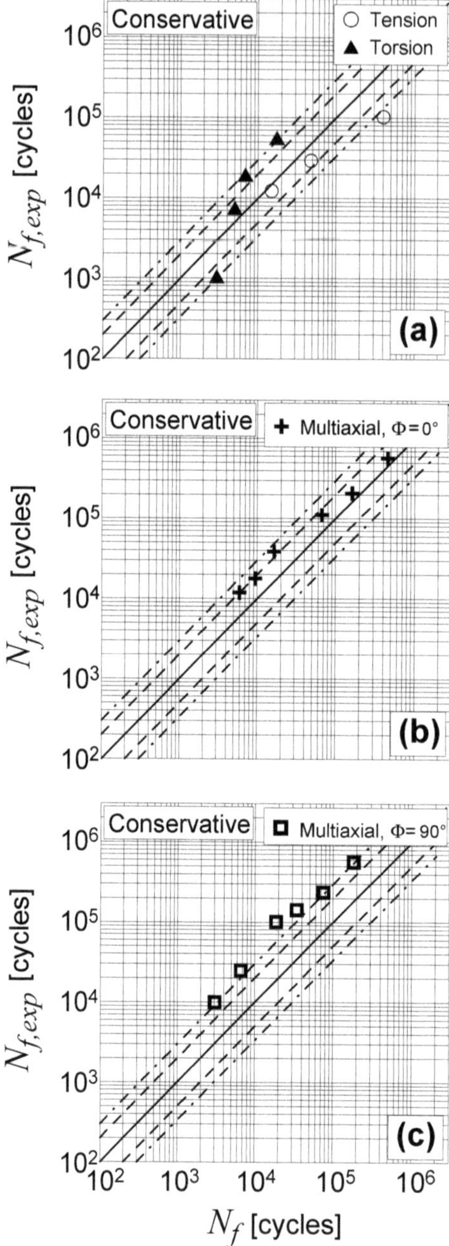

Figure 4. Accuracy of the present criterion in estimating the fatigue lifetime of Ti-6Al-4V notched specimens: (**a**) Uniaxial loading, (**b**) multiaxial proportional loading, (**c**) multiaxial non-proportional loading.

Figure 5 shows $N_{f,\text{exp}}$ as a function of the equivalent strain amplitude, $\varepsilon_{eq,\,a}$ (see Equation (4)). Note that the solid line is the experimental tensile Manson–Coffin equation. Since all the theoretical data lie very close to the experimental curve, it can be concluded that the accuracy level of the employed criterion is satisfactory.

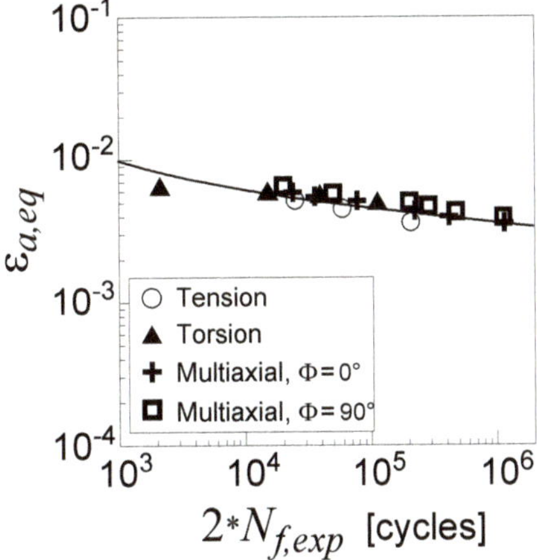

Figure 5. Experimental fatigue life $N_{f,\text{exp}}$ against equivalent strain amplitude $\varepsilon_{eq,a}$.

The above considerations can also be made by examining the values of the error index, I, computed as follows [24]:

$$I = \frac{\varepsilon_{eq,a} - \varepsilon_a}{\varepsilon_a} \cdot 100\% \tag{5}$$

In particular, Figure 6 shows the relative frequency of the I absolute value. It can be observed that the frequency distribution is close to zero, with 74% of the results in the range $0\% \leq |I| \leq 15\%$.

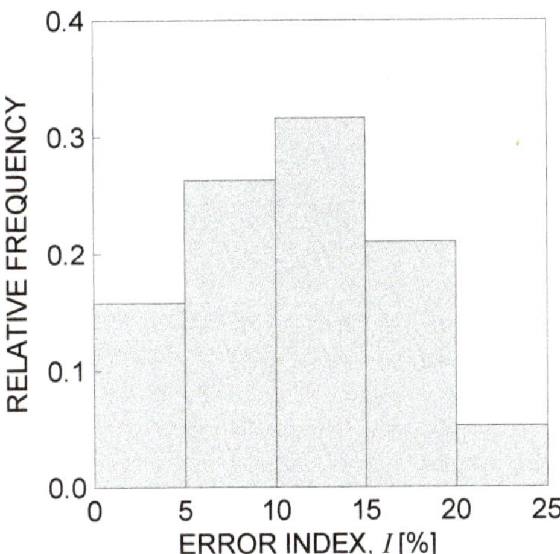

Figure 6. Absolute value of the error index I according to the present criterion.

In conclusion, we can remark that the implementation of energy concepts, based on the control volume, in the above strain-based multiaxial fatigue criterion appears an interesting tool for evaluating the fatigue behavior of severely notched components.

4. Conclusions

In the present paper, the fatigue life assessment of notched structural components has been performed by applying a critical plane-based multiaxial fatigue criterion. Such a criterion has been formulated by using the control volume concept related to the strain energy density criterion.

The material point located at a given distance from the notch tip is assumed to be the verification point. Such a distance has been taken to be a function of the control volume radii around the notch tip under both Mode I and Mode III loading. Once the position of the verification point and the orientation of the critical plane have been analytically determined, the fatigue lifetime has theoretically been evaluated through an equivalent normal strain amplitude, acting on the critical plane, together with the tensile Manson–Coffin curve.

The accuracy of the present criterion is evaluated through experimental data available in the literature, related to titanium alloy V-notched specimens under uniaxial and multiaxial fatigue loading. As far as the experimental data here examined are concerned, the joint application of the strain-based criterion and the control volume concept provides quite satisfactory fatigue life estimations.

On the basis of the encouraging results herein obtained, the present criterion seems to be able to correctly estimate the fatigue life of a structure with a stress concentrator (as a notch), by reaching an adequate margin of safety. However, different materials, notch geometries, and loading conditions need to be examined in order to develop a useful fatigue design tool.

Author Contributions: Conceptualization, A.C., C.R. and S.V.; methodology, A.C. and S.V.; validation, C.R. and S.V.; formal analysis, C.R.; writing—original draft preparation, C.R. and S.V.; writing—review and editing, A.C.; funding acquisition, A.C. and S.V.

Funding: This research was founded by the Italian Ministry for University and Technological and Scientific Research (MIUR), Research Grant PRIN 2015 No. 2015JW9NJT on "Advanced mechanical modeling of new materials and structures for the solution of 2020 Horizon challenges".

Acknowledgments: The authors gratefully acknowledge the financial support provided by the Italian Ministry for University and Technological and Scientific Research (MIUR), Research Grant PRIN 2015 No. 2015JW9NJT on "Advanced mechanical modeling of new materials and structures for the solution of 2020 Horizon challenges".

Conflicts of Interest: The authors declare no conflict of interest.

Nomenclature

$1, 2, 3$	principal strain directions
$\hat{1}, \hat{2}, \hat{3}$	averaged principal strain directions
E	elastic modulus
I	error index
N_f	theoretical fatigue life
$N_{f,exp}$	experimental fatigue life
P	verification point
r	distance of the verification point P from the notch tip
R_m	mean control volume radius
R_1	control volume radius related to Mode I
R_3	control volume radius related to Mode III
t	time
w	perpendicular unit vector to the critical plane
γ_a	Manson–Coffin shear strain amplitude
δ	angle between the averaged direction $\hat{1}$ and the normal w to the critical plane
ΔK_{1A}	notch stress intensity factor range under Mode I
ΔK_{3A}	notch stress intensity factor range under Mode III
$\Delta\sigma_{1A}$	high-cycle fatigue strength of smooth specimens under Mode I
$\Delta\tau_{3A}$	high-cycle Fatigue strength of smooth specimens under Mode III
$\varepsilon_1, \varepsilon_2, \varepsilon_3$	principal strains, with $\varepsilon_1 \geq \varepsilon_2 \geq \varepsilon_3$

ε_a	Manson–Coffin normal strain amplitude
$\varepsilon_{eq,a}$	equivalent normal strain amplitude
$\eta_{N,a}$	amplitude of the normal displacement vector component acting on the critical plane
$\eta_{C,a}$	amplitude of the tangential displacement vector component acting on the critical plane
λ	biaxiality ratio
ν_e	elastic Poisson ratio
ν_{eff}	effective Poisson ratio
ν_p	plastic Poisson ratio
σ_u	ultimate tensile strength
σ_y	yield strength
Φ	phase angle between tension and torsion loading

References

1. Jasper, T.M. The value of the energy relation in the testing of ferrous metals at varying ranges of stress and at intermediate and high temperatures. *Philos. Mag. Ser.* **1923**, *6*, 609–627. [CrossRef]
2. Macha, E.; Sonsino, C.M. Energy criteria of multiaxial fatigue failure. *Fatigue Fract. Eng. Mater. Struct.* **1999**, *22*, 1053–1070. [CrossRef]
3. Garud, Y.S. A new approach to the evaluation of fatigue under multiaxial loadings. *J. Eng. Mater. Technol.* **1981**, *103*, 118–125. [CrossRef]
4. Ellyin, F. Cyclic strain energy density as a criterion for multiaxial fatigue failure. In *Biaxial and Multiaxial Fatigue*; EGF, 3, Carpinteri, A., De Freitas, M., Spagnoli, A., Eds.; Mechanical Engineering: London, UK, 1989; pp. 571–583.
5. Ellyin, F.; Golos, K.; Xia, Z. In-phase and out-of-phase multiaxial fatigue. *J. Eng. Mater. Technol.* **1991**, *113*, 112–118. [CrossRef]
6. Ellyin, F.; Xia, Z. A general fatigue theory and its application to out-of-phase cyclic loading. *J. Eng. Mater. Technol.* **1993**, *115*, 411–416. [CrossRef]
7. Molsky, K.; Glinka, G. A method of elastic-plastic stress and strain calculation at a notch root. *Mater. Sci. Eng.* **1981**, *50*, 93–100. [CrossRef]
8. Glinka, G. Energy density approach to calculation of inelastic strain-stress near notches and cracks. *Eng. Fract. Mech.* **1985**, *22*, 485–508. [CrossRef]
9. Moftakhar, A.; Buczynski, A.; Glinka, G. Calculation of elasto-plastic strains and stresses in notches under multiaxial loading. *Int. J Fracture* **1995**, *70*, 357–373. [CrossRef]
10. Park, J.; Nelson, D. Evaluation of an energy-based approach and a critical plane approach for predicting constant amplitude multiaxial fatigue life. *Int. J. Fatigue* **2000**, *22*, 23–39. [CrossRef]
11. Lazzarin, P.; Zambardi, R. A finite-volume-energy based approach to predict the static and fatigue behavior of components with sharp V-shaped notches. *Int. J. Fract.* **2001**, *112*, 275–298. [CrossRef]
12. Lazzarin, P.; Berto, F. Some expressions for the strain energy in a finite volume surrounding the root of blunt V-notches. *Int. J. Fract.* **2005**, *135*, 161–185. [CrossRef]
13. Atzori, B.; Berto, F.; Lazzarin, P.; Quaresimin, M. Multi-axial fatigue behaviour of a severely notched carbon steel. *Int. J. Fatigue* **2006**, *28*, 485–493. [CrossRef]
14. Berto, F.; Lazzarin, P.; Tovo, R. Multiaxial fatigue strength of severely notched cast iron specimens. *Int. J. Fatigue* **2014**, *67*, 15–27. [CrossRef]
15. Meneghetti, G.; Campagnolo, A.; Berto, F.; Atzori, B. Averaged strain energy density evaluated rapidly from the singular peak stresses by FEM: cracked components under mixed-mode (I+II) loading. *Theor. Appl. Fract. Mech.* **2015**, *79*, 113–124. [CrossRef]
16. Ayatollahi, M.R.; Berto, F.; Campagnolo, A.; Gallo, P.; Tang, K. Review of local strain energy density theory for the fracture assessment of V-notches under mixed mode loading. *Eng. Solid Mech.* **2017**, *5*, 113–132. [CrossRef]
17. Berto, F.; Ayatollahi, M.R.; Campagnolo, A. Fracture tests under mixed mode I + III loading: An assessment based on the local energy. *Int. J. Damage Mech.* **2017**, *26*, 881–894. [CrossRef]
18. Torabi, A.R.; Berto, F.; Razavi, S.M.J. Ductile failure prediction of thin notched aluminum plates subjected to combined tension-shear loading. *Theor. Appl. Fract. Mech.* **2018**, *97*, 280–288. [CrossRef]

terms of plastic and elastic strain energy density (SED) at the weld seam, and relevant number of cycles in the SED method to estimate fatigue life of welded joints [11,12]. However, the selection of $S - N$ curves from different fatigue classes could be subjective, since they depend on the geometry of welded joints and loading modes. The corresponding mechanical responses are not easy to precisely determine as well. Another fracture mechanics approach could make the determination of lifetime for cracked welded components based on the Paris law, according to the relationship between the crack growth rate and stress intensity factor range [13]. Nevertheless, it is sensitive to the initial crack size, and thus the known crack is needed.

The intrinsic energy dissipation method could rapidly confirm the fatigue limit and evaluate life expectancy, based on the thermal evolution on the surface of welded joints under cyclic loadings obtained from the thermo-graphic measurement [14–16]. This approach is independent of geometrical characteristics of welded joints, and does not require some given assumptions or prerequisites, compared with the notch stress approach, fracture mechanics method, and so on. This temperature variation is representative of the microstructure's movement and evolution in some way, which could be deemed as the connection between different scales. In the meantime, the intrinsic dissipation method supported by the thermo-graphic technique has great advantages in saving time and specimens. It could rapidly confirm the high-cycle fatigue limit, and construct a fatigue life assessment model via a very limited number of tests without sacrificing the calculation precision, proposed by Risitano and Luong et al. [17,18]. Essentially, the dissipation of energy directly reflects a material's fatigue failure mechanism, caused by the internal friction between the micro-structures. Meanwhile, the direction judgment falls into disuse when the dissipated energy is treated as damage parameter. Currently, a so-called Quantitative Thermo-graphic Methodology (QTM) has been widely used in studying the fatigue behavior of pure metal materials, notched specimens, and welded joints [19,20]. However, this approach is based on the assumption of a constant temperature increment during the stabilized stage [21,22], and the temperature is directly considered a fatigue damage parameter. In this paper, the linear increase of temperature increments in the stable stage II was measured on the surface of a flat butt welded joint, and was taken into account for constructing a fatigue life estimation model of welded joints. A well-suited energy form (instead of temperature) was treated as the fatigue indicator for characterizing the heat dissipation behavior.

First, in order to eliminate the thermal elastic effect caused by the welding residual stress, annealing was carried out. Its mechanical behavior was obtained through a monotonic test. The fatigue tests were conducted on the butt joint under fully reversed tension-compression loading. Then, based on the fatigue life estimations model, taking the linear temperature evolution and the intrinsic dissipation into account, the estimated life spans calculated from the energy-based approach were compared with experimental ones.

2. Experimental Work

2.1. Specimen Preparation

The flat butt-welded joints were designed, as in Figure 1, and connected by the arc-welding process. The weld seam is located at the middle of the specimen (purple marks). Its thickness is 6 mm. The parental material is high strength steel, and its yield strength is 460 MPa, which is widely utilized in bearing structures, like bridges or buildings.

Some research [19,20] found that if a mean stress existed, there would be a very small heat contribution offered by a thermo-elastic heat source. In general, the welding procedure could produce residual stress located at the welded zone. This tensile residual stress is usually treated as a kind of equivalent mean stress. In order to eliminate the inherent mean stress, the specimens were annealed by raising the temperature to 500 °C and preserving heat for two hours. After they were cooled to room temperature, the residual stresses on the surface of the specimen around the welded area were measured by X-ray, as shown in Figure 2. The three measuring points are shown in Figure 3, and the tested results are listed in Table 1. The residual stresses of those points on the surface of the specimen are all small, and could be considered as non-residual stress.

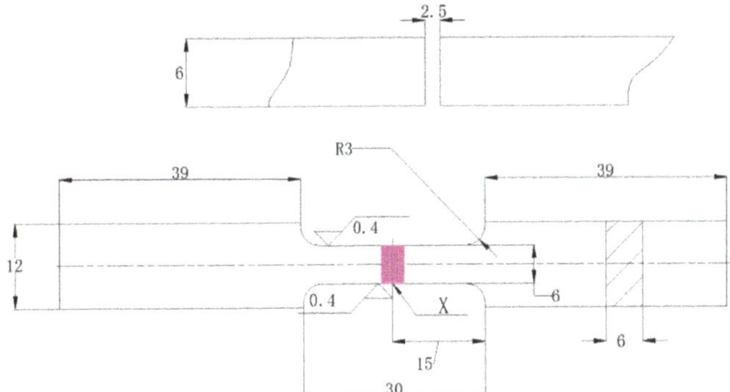

Figure 1. Specimen geometry of welded joint (all dimensions in mm, roughness in micrometers, tolerance for all dimensions ±0.01).

Figure 2. Residual stress measurement by X-ray.

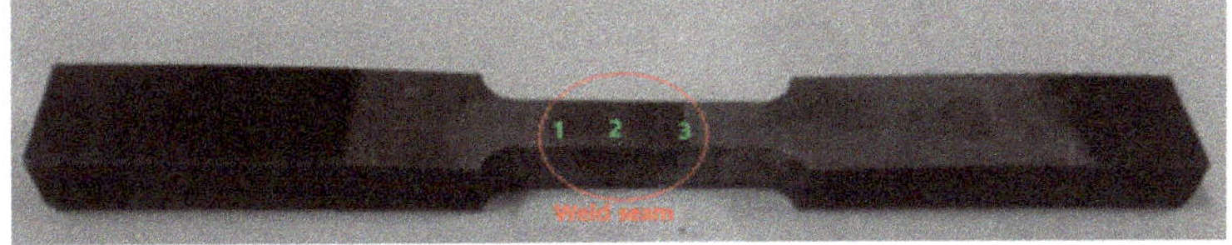

Figure 3. Layout of measurement points.

Table 1. Residual stress measured by X-ray.

Point 1 (MPa)	Point 2 (MPa)	Point 3 (MPa)
-14.55 ± 3.27	-16.84 ± 12.51	5.09 ± 6.86

2.2. Monotonic Test

A monotonic tensile test was conducted on the butt joint specimen under displacement control at the rate of 0.01 mm/s. An Instron uniaxial extensometer (INSTRON, Norwood, MA, USA) with a 12 mm gauge length was utilized for measuring the strain. The true and engineering stress–strain curves are both shown in Figure 4. According to the stress and strain responses, the mechanical parameters of welded joints are listed in Table 2. The yield strength of welded joints is less than the parental material, which is possibly caused by the welding process parameters and the welding wire material.

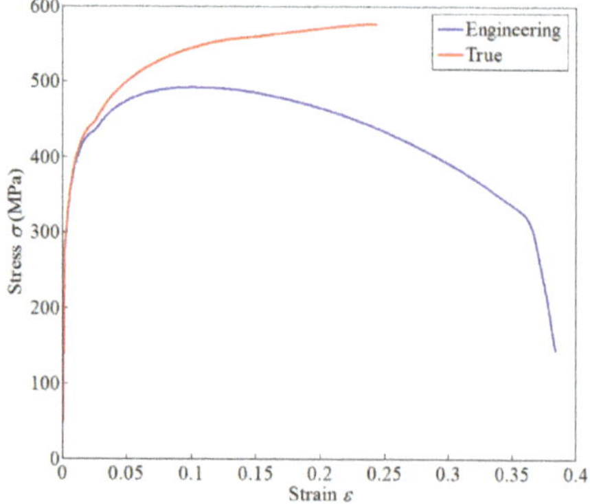

Figure 4. Stress and strain responses under monotonic tensile loading.

Table 2. Mechanical parameters of welded joint.

Young's Modulus E (GPa)	205
Yield Strength σ_y (MPa)	365
Ultimate Tensile Strength σ_{UTS} (MPa)	500

2.3. Cyclic Test

The installation diagram of the uniaxial fatigue test and thermal measurement is shown in Figure 5. The temperature distribution contour on the surface of the specimen during the cyclic loading was recorded by a high-performance infrared (IR) camera (FLIR, Boston, MA, USA), as shown in Figure 6. This camera has high sensitivity to perceive the thermal variation of 20 mK at room temperature, and its resolution could reach 640×512 pixels in full field mode.

In the fully reversed fatigue tests, the applied constant stress amplitudes were 240 MPa, 260 MPa, 300 MPa and 340 MPa, which aimed at constructing a fatigue life estimation model and testing its prediction accuracy. All the fatigue tests were under load control, and the failure criterion was 50% drop of displacement. The frequency of applied uniaxial loading was 20 Hz.

Figure 5. Installation diagram of the fatigue test and thermal measurement.

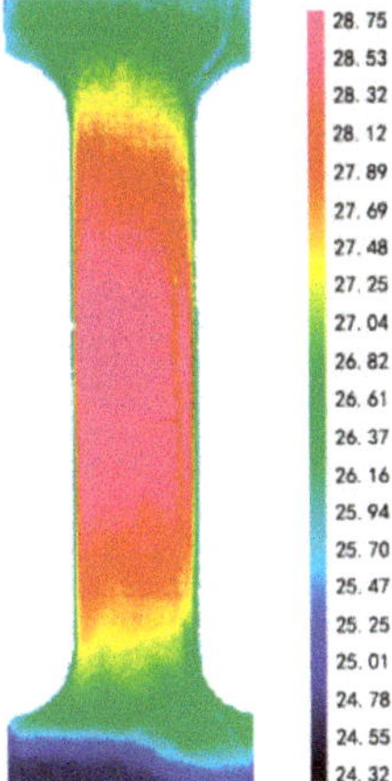

Figure 6. Temperature distribution contour measured by IR camera (unit: °C).

2.4. Thermal Evolution

The heat caused by the internal friction between microstructures under cyclic loading is dissipated through weld metal. The temperature increment θ with respect to initial temperature could be obtained from the thermal-graphic measurement. The thermal evolution in whole life cycles under stress amplitudes 300 MPa and 260 MPa are shown in Figures 7 and 8, respectively. Before the temperature increment comes into the stable stage II, as shown in Figure 7, the temperature of the material surface increases continuously, due to cyclic straining, until a relative equilibrium state between the internal heat production and external heat exchange is reached [19–23]. Being different from pure metal materials, the temperature increment of this welded joint linearly increases during the stable stage II both in high stress amplitude and low stress amplitude. Thus, the method for calculating accumulated intrinsic dissipation and predicting lifespan has to be developed.

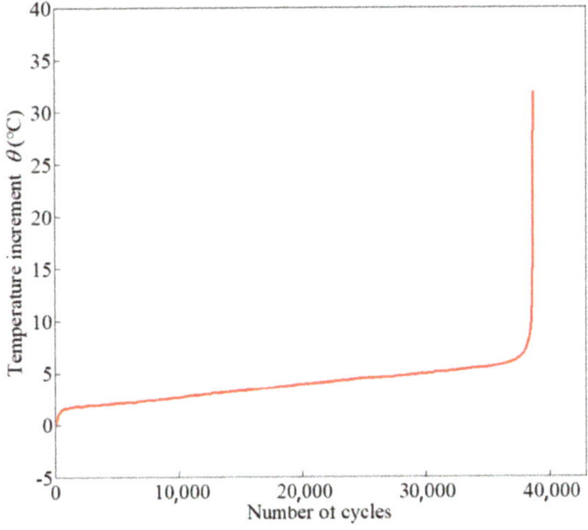

Figure 7. The thermal evolution in whole life cycles under stress amplitude 300 MPa.

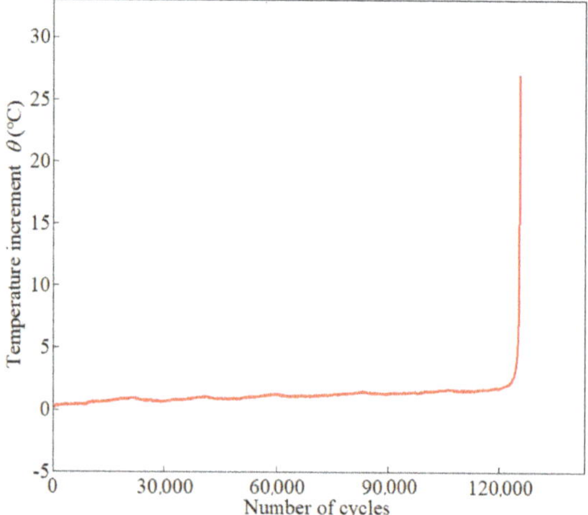

Figure 8. The thermal evolution in whole life cycles under stress amplitude 260 MPa.

3. An Energy-Based Fatigue Life Prediction Method

3.1. Intrinsic Dissipation

According to the literature [23,24], the intrinsic dissipation based on a quantitative thermo-graphic method is considered as the fatigue indicator, instead of temperature produced on the material surface under cyclic loads. From the perspective of the fatigue damage mechanism, repeated work generated by the external loads facilitates micro grain movement and friction until the macro crack appears. During this damage evolution process, the mechanical energy transforms into heat energy, elastic and plastic strain energy, storage energy, and so on. Certainly, most of them are released in the form of heat energy, which results in thermal variations in the procedure of heat conduction, heat exchange, heat convection and heat radiation. However, it is not hard to see that the intrinsic dissipation determines thermal evolution, which is more suitable to describe fatigue failure behavior. It is assumed that there is an external heat source r during cyclic loading, while the total heat source s mainly includes three parts:

$$s = s_{the} + d_1 + r \tag{1}$$

where s_{the} is the thermoelastic heat source generated by the reversible elastic strain, and d_1 represents the intrinsic dissipation caused by the irreversible strains.

In view of thermodynamic laws, the heat source could be expressed by temperature differential equation, presented in the previous references [20–24]:

$$s = \rho C \frac{\partial \theta}{\partial t} - k\Delta\theta \tag{2}$$

where ρ is density, C is specific heat, k is isotropic thermal conductivity, θ is temperature increment with respect to the initial temperature, t is time, and Δ is the Laplace operator. According to Chrysochoos and co-authors' work [25], proposing 0D, 1D, 2D methods to solve the above differential equation, the 0D model is considered as the most appropriate approach for the heat solution in the high cycle fatigue test. Then, the term $k\Delta\theta$ could be removed from Equation (2).

When the stress ratio R is minus one, the heat source s_{the} tends to be zero because of the thermo-elastic coupling effect under tension-compression loading. If there is a mean stress, the heat contribution provided by the heat source s_{the} is extremely small, so that it could be neglected in the thermal differential equation. The external heat source r is defined to describe heat conversion between the specimen and the environment. However, since the thermal parameters and exchange conditions are not easy to accurately determine, it has to be simplified according to the heat loss behavior. Finally, a linear equation is utilized to establish the relationship between the heat loss and temperature variation:

$$r = -\rho C \frac{\theta}{\tau} \tag{3}$$

where τ stands for a time constant characterizing the heat loss.

After Equation (1), Equation (2) and Equation (3) are integrated and deduced, the intrinsic dissipation d_1 could be expressed like this:

$$d_1 = \rho C \left(\frac{\partial \theta}{\partial t} + \frac{\theta}{\tau} \right) \tag{4}$$

It is worth mentioning that the intrinsic dissipation d_1 is defined as the average dissipated energy intensity of the affected region on the surface of the testing specimen. The heat diffusion along the thickness of specimen is assumed to be identical, while the heat loss between the specimen and the environment is still characterized by the term $\rho C \theta / \tau$, as shown in Equation (4).

3.2. Energy-Based Fatigue Life Model

In this paper, the intrinsic dissipation d_1 is regarded as the fatigue indicator. Once the fatigue damage is produced by each cyclic load, the dissipated energy is accumulated. In order to simplify calculation of the accumulated intrinsic dissipation d_{1c}, the time t could be converted into the term N/f ($f = N/t$). Then, the total accumulated intrinsic dissipation d_{1c} could be obtained from the definite integral of the intrinsic dissipation d_1 during the whole life cycle:

$$d_{1c} = \int_0^{N_f} d_1 / f \, dN \tag{5}$$

Based on the damage evolution law, the total accumulated intrinsic dissipation d_{1c} has clear physical meaning, which represents the energy tolerance to failure of a kind of material applied by the cyclic loading. In other words, it is equal to the energy tolerance to failure E_c.

In the Equation (4), it is noticed that the term $\partial \theta / \partial t$ would be zero if the asymptotic portion of temperature increase in stage II θ_{AS} was constant. A large number of studies [23,24] support this conclusion for most pure materials. However, the welding process brings impurities and porosities, which results in non-linear fatigue damage evolution behavior and varying thermal dissipation during the whole fatigue process. Some researchers [22–27] also found there is the linear temperature evolution after the material gets into a stabilized stage II, as shown in Figure 9. Then, the energy tolerance to failure E_c could be evaluated like this:

$$E_c = d_{1c} = \int_0^{N_f} \frac{\rho C}{f} \left(\frac{\partial \theta}{\partial N} + \frac{\theta}{\tau} \right) dN \tag{6}$$

Wang et al. [28] presented that the temperature increment θ in stage II could be converted into the asymptotic portion of temperature increase θ_{AS} and the incremental temperature θ_Δ. It should be noted that the incremental temperature θ_Δ represents the increased temperature with respect to the asymptotic portion of temperature increase θ_{AS} during the stabilized stage II. According to the linear evolution behavior of the temperature increment θ in stage II versus the life cycles, the slope of the temperature variation curve is treated as the temperature increment rate λ, which is expressed like this:

$$\lambda = \frac{\partial \theta}{\partial N} = \frac{\theta_\Delta}{N} \tag{7}$$

Then, substituting Equation (7) into Equation (6) and taking the decomposition terms θ_{AS} and θ_Δ into account, the energy tolerance to failure E_c is resolved by the following equation:

$$E_c = \int_0^{N_f} \frac{\rho C}{f}\left(\frac{\partial \theta}{\partial N} + \frac{\theta_{AS} + \lambda N}{\tau}\right) dN \tag{8}$$

In fact, the number of cycles in both initial stage I and final stage III only accounts for a small part of whole life cycles, and could be ignored in the following definite integral of Equation (8). Then, its integrating result is indicated like this:

$$E_c = \frac{\rho C \lambda}{2f\tau} N_f^2 + \frac{\rho C \lambda}{f} N_f + \frac{\rho C \theta_{AS}}{f\tau} N_f \tag{9}$$

According to Equation (9), the fatigue life N_f related with dissipated energy could be calculated as the following equation:

$$N_f = \frac{f\tau}{\rho C \lambda}\left(\sqrt{2\frac{\rho C \lambda}{f\tau} E_c + \left(\frac{\rho C}{f}\lambda + \frac{\rho C \theta_{AS}}{f\tau}\right)^2} - \frac{\rho C}{f}\lambda - \frac{\rho C \theta_{AS}}{f\tau}\right) \tag{10}$$

Based on Equation (10), the cyclic loading had to at least reach into the early stage II for calculating the asymptotic portion of temperature increase θ_{AS} and the temperature increment rate λ. Once these two parameters were defined, the remaining lifetime or fatigue damage could be derived according to Equation (10) regardless of whether the thermal measurement is suspended or not.

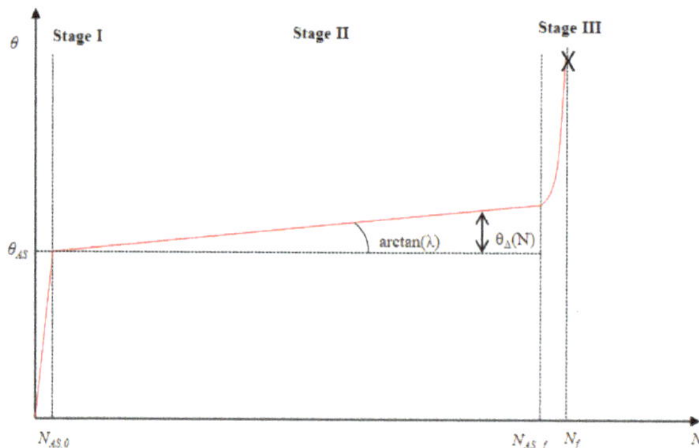

Figure 9. Schematic diagram of temperature evolution versus life cycles during three stages.

In order to calculate intrinsic dissipation, the thermo-physical parameters and loading frequency needed to calculate the intrinsic dissipation are given in Table 3. It is worth mentioning that the time parameter τ was ascertained by calculating the heat loss from the peak temperature to the room temperature, also listed in Table 3.

Table 3. Thermo-physical and boundary parameters of the welded joints.

ρ (kg·m^{-3})	C (J·kg^{-1}·K^{-1})	k (WK^{-1}·m^{-1})	τ (s)	f (Hz)
7850	460	45.5	14.7~35.2	20

Furthermore, from the above Equation (9), the energy tolerance to failure E_c could be estimated once the fatigue life and loading conditions are determined. Combined, the thermal evolution in the whole lifespan under constant stress amplitudes with thermo-physical parameters, and the energy tolerances to failure E_c for different stress amplitudes are listed in Table 4. It is very interesting to note that all energy tolerances to failure E_c are close to 1.0×10^9 J/m^3. The energy tolerance to failure E_c represents the inherent nature of the material or structure in some way and is independent of loading conditions. Its significance lies in allowable precise fatigue life estimation, based on Equation (10), if the thermal evolution within a certain cycle could be ascertained.

Table 4. Energy tolerance to failure E_C under different stress amplitude.

Stress Ratio R	Stress Amplitude σ_a (MPa)	E_C (J/m^3)
-1	300	1.0091×10^9
-1	260	0.9653×10^9

Based on the results mentioned above, the thermal evolution in 2000 cycles under stress amplitudes 340 MPa and 240 MPa are shown in Figures 10 and 11, respectively. The asymptotic portion of temperature increase θ_{AS} during the stabilized stage II and the temperature increment rate λ could be obtained from the temperature variation. Then, according to Equation (10), their life expectancies are assessed and listed in Table 5. The predicted cycles for these welded joints under high stress amplitude and low high stress amplitude are both close to experimental ones. This energy-based approach could effectively predict fatigue life for high strength welded joints through thermal-graphic measurement, merely based on the thermal evolution in several thousand cycles. Once the energy tolerance to failure E_c is determined, only a very small amount of time is needed to predict life expectancy. Notably, this specimen is a typical flat butt welded joint and is applied with traditional fully reversed cyclic loadings, so its thermal evolution behavior could be helpful for characterizing the energy dissipation mechanism of similar welded structures with the same loading modes.

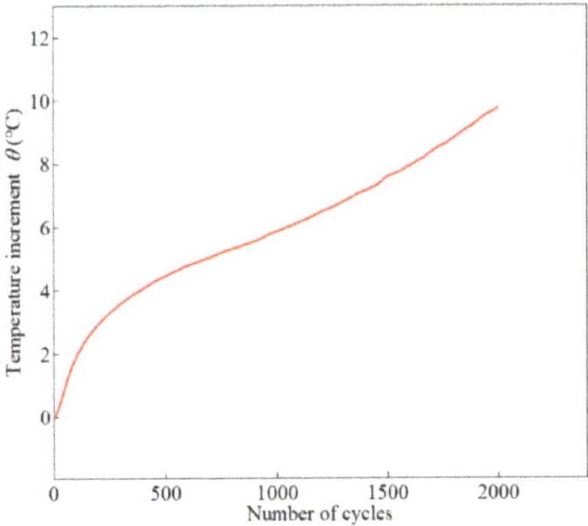

Figure 10. The thermal evolution in 2000 cycles under stress amplitude 340 MPa.

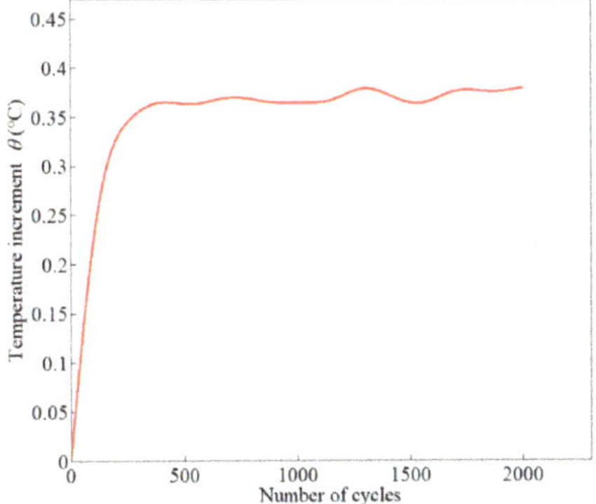

Figure 11. The thermal evolution in 2000 cycles under stress amplitude 240 MPa.

Table 5. Predicted results under different stress amplitude.

Stress Ratio R	Stress Amplitude σ_a (MPa)	Predicted Cycles	Experimental Cycles
−1	340	11,809	12,877
−1	240	401,025	386,302

4. Conclusions

A fatigue life assessment model based on intrinsic energy dissipation was conducted on high strength steel welded joints. This method suggested that the intrinsic energy dissipation was considered as a fatigue indicator, rather than temperature. In this model, the linear temperature evolution in stage II was fully taken into account to guarantee calculation precision, and its predicted results were in good agreement with experimental ones for different stress amplitudes. This approach had an apparent distinct advantage in testing efficiency, because limited testing time and specimens were needed for lifetime modeling and estimation.

It was interesting to find that the energy tolerance to failure E_c tended to be a constant for these welded joints under different stress amplitudes. In some way, this parameter characterized fatigue damage accumulation in the whole lifespan, and was equal and load-independent. This result may also be helpful for predicting the remaining service life of welded engineering structures, once the thermal evolution in certain cycles is obtained.

Author Contributions: Conceptualization, C.M. and F.B.; Methodology, C.M.; Software, W.L.; Validation, W.L.; Formal Analysis, C.M.; Investigation, C.M.; Resources, C.M.; Data Curation, X.X.; Writing-Original Draft Preparation, C.M.; Writing-Review & Editing, F.B.; Visualization, X.X.; Supervision, F.B.; Project Administration, C.M.; Funding Acquisition, C.M.

Funding: This research was funded by [Natural Science Foundation of Hunan Province] grant number [2017JJ3059], and [Postdoctoral Science Foundation] grant number [2017M622569], and [State Key Laboratory of Advanced Design and Manufacturing for Vehicle Body] grant number [31715012].

Acknowledgments: The authors gratefully thank for the support of the Hunan Province Natural Science Foundation (Grant No.: 2017JJ3059) and China Postdoctoral Science Foundation (Grant No.: 2017M622569) and Open Foundation of State Key Laboratory of Advanced Design and Manufacturing for Vehicle Body (Grant No.: 31715012).

Conflicts of Interest: The authors declare no conflict of interest. The funders had no role in the design of the study; in the collection, analyses, or interpretation of data; in the writing of the manuscript, and in the decision to publish the results.

References

1. Shiozaki, T.; Yamaguchi, N.; Tamai, Y.; Hiramoto, J.; Ogawa, K. Effect of weld toe geometry on fatigue life of lap fillet welded ultra-high strength steel joints. *Int. J. Fatigue* **2018**, *116*, 409–420. [CrossRef]
2. Costa, J.D.M.; Ferreira, J.A.M.; Abreu, L.P.M. Fatigue behavior of butt welded joints in a high strength steel. *Procedia Eng.* **2010**, *2*, 697–705. [CrossRef]
3. Ma, L.L.; Wei, Y.H.; Hou, L.F.; Guo, C.L. Evaluation on fatigue performance and fracture mechanism of laser welded TWIP steel joint based on evolution of microstructure and micromechanical properties. *J. Iron Steel Res. Int.* **2016**, *23*, 677–684. [CrossRef]
4. D'Angelo, L.; Nussbaumer, A. Estimation of fatigue S-N curves of welded joints using advanced probabilistic approach. *Int. J. Fatigue* **2017**, *97*, 98–113. [CrossRef]
5. Ding, Y.L.; Song, Y.S.; Cao, B.Y.; Wang, G.X.; Li, A.Q. Full-range S-N fatigue life evaluation method for welded bridge structures considering hot-spot and welding residual stress. *J. Bridge Eng.* **2016**, *21*, 1–10. [CrossRef]
6. Deng, C.; Liu, Y.; Gong, B.; Wang, D. Numerical implementation for fatigue assessment of butt joint improved by high frequency mechanical impact treatment: A structural hot spot stress approach. *Int. J. Fatigue* **2016**, *92*, 211–219. [CrossRef]
7. Stenberg, T.; Barsoim, Z.; Balawi, S.O.M. Comparison of local stress based concepts- Effects of low- and high cycle fatigue and weld quality. *Eng. Fail. Anal.* **2015**, *57*, 323–333. [CrossRef]

8. Schork, B.; Kucharczyk, P.; Madia, M.; Zerbst, U.; Hensel, J.; Bernhard, J.; Tchuindjang, D.; Kaffenberger, M.; Oechsner, M. The effect of the local and global weld geometry as well as material defects on crack initiation and fatigue strength. *Eng. Fract. Mech.* **2018**, *198*, 103–122. [CrossRef]

9. Varvani-Farahani, A.; Kodric, T.; Ghahramani, A. A method of fatigue prediction in notched and un-notched components. *J. Mater. Process. Technol.* **2005**, *169*, 94–102. [CrossRef]

10. Nykanen, T.; Mettanen, H.; Bjork, T.; Ahola, A. Fatigue assessment of welded joints under variable amplitude loading using a novel notch stress approach. *Int. J. Fatigue* **2017**, *101*, 177–191. [CrossRef]

11. Gu, Z.; Mi, C.; Ding, Z.; Zhang, Y.; Liu, S.; Nie, D. An energy-based fatigue life prediction of a mining truck welded frame. *J. Mech. Sci. Technol.* **2016**, *30*, 3615–3624. [CrossRef]

12. Berto, F.; Vinogradov, A.; Filippi, S. Application of the strain energy density approach in comparing different design solutions for improving the fatigue strength of load carrying shear welded joints. *Int. J. Fatigue* **2017**, *101*, 371–384. [CrossRef]

13. Chapetti, M.D.; Jaureguizahar, L.F. Fatigue behavior prediction of welded joints by using an integrated fracture mechanics approach. *Int. J. Fatigue* **2012**, *43*, 43–53. [CrossRef]

14. Meneghetti, G.; Ricotta, M. The use of the specific heat loss to analyze the low- and high-cycle fatigue behavior of plain and notched specimens made of a stainless steel. *Eng. Fract. Mech.* **2012**, *81*, 2–16. [CrossRef]

15. Palumbo, D.; Galietti, U. Characterisation of steel welded joints by infrared thermographic methods. *Quant. Infrared Thermogr. J.* **2014**, *11*, 29–42. [CrossRef]

16. Palumbo, D.; de Finis, R.; Ancona, F.; Galietti, U. Damage monitoring in fracture mechanics by evaluation of the heat dissipated in the cyclic plastic zone ahead of the crack tip with thermal measurement. *Eng. Fract. Mech.* **2017**, *181*, 65–76. [CrossRef]

17. Luong, M.P. Fatigue limit evaluation of metals using an infrared thermographic technique. *Mech. Mater.* **1998**, *28*, 155–163. [CrossRef]

18. La Roa, G.; Risitano, A. Thermographic methodology for rapid determination of the fatigue limit of materials and mechanical components. *Int. J. Fatigue* **2000**, *22*, 65–73. [CrossRef]

19. Wang, X.G.; Crupi, V.; Jiang, C.; Guglielmino, E. Quantitative thermographic methodology for fatigue life assessment in a multiscale energy dissipation framework. *Int. J. Fatigue* **2015**, *81*, 249–256. [CrossRef]

20. Wang, X.G.; Crupi, V.; Guo, X.L.; Zhao, Y.G. Quantitative thermographic methodology for fatigue assessment and stress measurement. *Int. J. Fatigue* **2010**, *32*, 1970–1976. [CrossRef]

21. Corigliano, P.; Crupi, V.; Epasto, G.; Guglielmino, E.; Risitano, G. Fatigue life prediction of high strength steel welded joints by energy approach. *Procedia Struct. Integr.* **2016**, *2*, 2156–2163. [CrossRef]

22. Crupi, V.; Gugliemino, E.; Maestro, M.; Marino, A. Fatigue analysis of butt welded AH36 steel joints: Thermographic method and design $S - N$ curve. *Mar. Struct.* **2009**, *22*, 373–386. [CrossRef]

23. Guo, Q.; Guo, X.; Fan, J.; Syed, R.; Wu, C. An energy method for rapid evaluation of high-cycle fatigue parameters based on intrinsic dissipation. *Int. J. Fatigue* **2015**, *80*, 136–144. [CrossRef]

24. Amiri, M.; Khonsari, M.M. Rapid determination of fatigue failure based on temperature evaluation: Fully reversed bending load. *Int. J. Fatigue* **2010**, *32*, 382–389. [CrossRef]

25. Berthel, B.; Chrysochoos, A.; Wattrisse, B.; Galtier, A. Infrared image processing for the calorimetric analysis of fatigue phenomena. *Exp. Mech.* **2008**, *48*, 79–90. [CrossRef]

26. Plekhov, O.A.; Saintier, N.; Palin-Luc, T.; Uvarov, S.V.; Naimark, O.B. Theoretical analysis, infrared and structural investigations of energy dissipation in metals under cyclic loading. *Mater. Sci. Eng. A* **2007**, *462*, 367–369. [CrossRef]

27. Dumoulin, S.; Louche, H.; Hopperstad, O.S.; Børvik, T. Heat sources, energy storage and dissipation in high-strength steels: Experiments and modeling. *Eur. J. Mech.* **2010**, *29*, 461–474. [CrossRef]

28. Wang, X.G.; Crupi, V.; Jiang, C.; Feng, E.S.; Guglielmino, E.; Wang, C.S. Energy-based approach for fatigue life prediction of pure copper. *Int. J. Fatigue* **2017**, *104*, 243–250. [CrossRef]

© 2019 by the authors. Licensee MDPI, Basel, Switzerland. This article is an open access article distributed under the terms and conditions of the Creative Commons Attribution (CC BY) license (http://creativecommons.org/licenses/by/4.0/).

Article

Multiscale Damage Evolution Analysis of Aluminum Alloy Based on Defect Visualization

Yuquan Bao [1], Yali Yang [1,*], Hao Chen [1,*], Yongfang Li [1], Jie Shen [2] and Shuwei Yang [1]

[1] School of Mechanical and Automotive Engineering, Shanghai University of Engineering Science, Shanghai 201620, China; m15221623776@163.com (Y.B.); liyongfang@sues.edu.cn (Y.L.); 18016334293@163.com (S.Y.)

[2] College of Engineering and Computer Science, University of Michigan, Dearborn, MI 48128, USA; shen@umich.edu

* Correspondence: carolyn71@163.com (Y.Y.); pschenhao@163.com (H.C.); Tel.: +86-135-8590-1281 (Y.Y.); +86-135-8590-1312 (H.C.)

Received: 29 October 2019; Accepted: 29 November 2019; Published: 3 December 2019

Featured Application: Life prediction for engineering materials.

Abstract: The evaluation of fatigue life through the mechanism of fatigue damage accumulation is still a challenging task in engineering structure failure analysis. A multiscale fatigue damage evolution model was proposed for describing both the mesoscopic voids propagation in the mesoscopic-scale and fatigue damage evolution process, reflecting the progressive degradation of metal components in the macro-scale. An effective method of defect classification was used to implement 3D reconstruction technology based on the MCT (micro-computed tomography) scanning damage data with ABAQUS subroutine. The effectiveness was validated through the comparison with the experimental data of fatigue damage accumulation. Our results indicated that the multiscale fatigue damage evolution model built a bridge between mesoscopic damage and macroscopic fracture, which not only used the damage variable in the macro-scale to characterize the mesoscopic damage evolution indirectly but also understood macroscopic material degradation behavior from mesoscale with sufficient precision. Furthermore, the multiscale fatigue damage evolution model could offer a new reasonable explanation of the effect of load sequence on fatigue life, and also could predict the fatigue life based on damage data by nondestructive testing techniques.

Keywords: multiscale; fatigue damage evolution; ABAQUS subroutine; 3D reconstruction; MCT scanning; fatigue life

---

1. Introduction

Fatigue fracture is one of the most common failure modes for engineering structures, and 80–90% of failures fall into this category [1–3]. Therefore, fatigue failure of metals has been the subject of study by many researchers. However, the understanding of fatigue mechanisms for evaluation of fatigue damage accumulation and fatigue life is still a challenging task up to now. With the development of theory and framework of the continuum damage mechanics (CDM), probably first presented by Kachanov [4], and the advance in technique for micro-observation of interior structure of steel materials, such as scanning electron microscopy [5], different methods and theories have been employed to study the evolution law of fatigue damage processes, such as the number of cycle load to failure, dissipated energy, and degradation in mechanical properties [6–16]. In recent years, more and more scholars have devoted to the study of fatigue damage processes.

On the one hand, many researchers pay attention to the continuous average damage variable to describe the degradation of material on a larger macro-scale, which is easy for engineering applications due to its simplicity and effectiveness. Since Miner expressed this concept in the description of fatigue damage accumulation in 1945 [17], the cumulative fatigue damage theories have been developed increasingly, including the works of Marco and Starkey [18], Henry [19], Gatts [20], Manson [21], Chaboche [22], and many others. As a result, many different fatigue damage models have been developed based on the concept of CDM developed by Chaboche, Manson [23], Franke [24], and many others. However, most of the methods for macroscopic fracture are short for a deep and comprehensive understanding of microscopic damage mechanics theory in the study of the macro-scale fatigue process.

On the other hand, the methods of microscopic fatigue analysis, which are used to describe the microscopic defects initiation and growth behavior, are studied by some researchers, e.g., Miller [25], Angelova [26], and Polák [27], among many others. McClintock [28] and Rice and Tracey [29] have studied the nucleation, growth, and coalescence of cylindrical and spherical voids related to the fracture theory. McDowell et al. [30] developed a micro-scale fatigue model, which was able to characterize the effect of many micro-structural entities (cracks, voids, grains, etc.) on high cycle fatigue response of metallic materials. Leuders et al. [31] used the computed tomography for detecting the distribution of voids in Ti–6Al–4V samples manufactured by selective laser melting. More recently, Hu [32] analyzed the fatigue crack growth process in the material by using synchrotron X-ray micro-computed tomography. However, such attempts providing insight into the fatigue damage process have built-in complexities, which require accurate and detailed knowledge of microstructural features.

Both of the above research methods are based on the study of fatigue damage evolution in a single macro-scale or microscale. On the macro-scale, using a macro-variable to describe the continuous average fatigue damage accumulation extent cannot be explained by the fatigue failure mechanisms viewed from a smaller microscale. Nevertheless, the micro-analysis of microscopic defect behavior, focusing on only a few single cracks or voids, cannot evaluate average fatigue damage extent in a larger macro-scale. It has long been regarded as a very important issue to establish the micro-macro relationship for the fatigue accumulation process because such a relationship can enhance our understanding of the fundamental nature of fatigue mechanisms, but it is still a challenging task up to now.

There are few scholars who have studied multiscale fatigue damage evolution process and models. For instance, Desmorat et al. [33] employed the Eshelby–Kröner scale transition law and used a double-scale model to describe the HCF (high-cycle fatigue) phenomenon in which damage occurred only at the microscopic scale. Wan et al. [34] considered the phenomenon of building orientations and porosity in the additive manufacture structures and used the micro-scale damage-evolution equation to describe the damage evolution process at the macroscopic scale. However, such attempts focused on exploring the evolution process of micro-defect size and neglected the influence of microscopic defect shape and position distribution on fatigue damage. There is a paucity of modeling methods and fatigue damage evolution analyses based on micro-defects visualization processing and 3D reconstruction, considering aspects of the real random uniform micro-structural morphology. In this regard, this paper aimed to create a multiscale damage evolution model by an efficient and simple defect classification method and 3D reconstruction technology based on MCT (micro-computed tomography) scanning data. Firstly, fatigue specimens at different loading stages were scanned by MCT technology. The defect information for fatigue specimens was graded and simplified by the defect classification method, considering not only the micro-defect size but also the shape and position distribution on fatigue damage via AVIZO (3D visualization software, September 2, 2016, FEI SAS, Mérignac, France) visual processing. Then, 3D reconstruction was carried out. An equivalent simplified model was established by the ABAQUS subroutine. This model provided an effective tool to build a bridge between mesoscopic damage and macroscopic fracture, using the damage variable in macro-scale to characterize the mesoscopic damage evolution indirectly. At the same time, the residual fatigue life for engineering structures under unknown loading times could be predicted easily through the microstructure by nondestructive detection based on this methodological study.

2. Experiment

2.1. Material and Specimen Preparation

The composition of 6061-T6 aluminum alloy, which was used to manufacture test specimens, is shown in Table 1.

Table 1. Composition of 6061-T6 aluminum alloy.

Chemical Composition	Cu	Si	Fe	Mn	Mg	Zn	Cr	Ti	Other	AL
Ratio	25%	60%	70%	15%	85%	25%	16%	15%	15%	margin

The test pieces were prepared in accordance with the American Society for Testing Materials ASTM E8/E8M-15a standard and the metal material axial equal amplitude cyclic fatigue test method. While high-speed cutting is required, the alignment of the specimen should be ensured. The surface roughness was controlled by a polishing treatment. The specimen designed in this paper is shown in Figure 1.

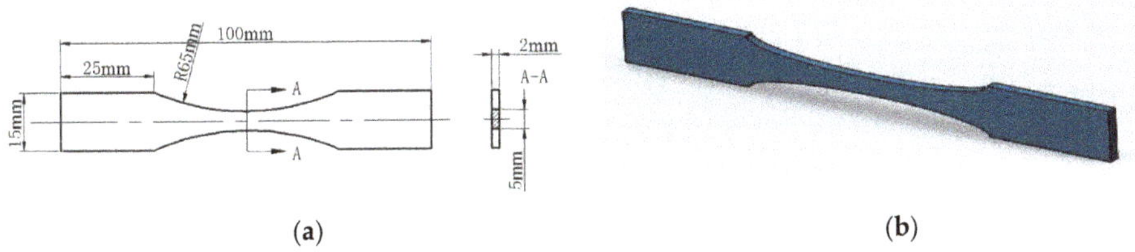

(a)   (b)

Figure 1. Specimens (**a**) Two-dimensions size of the specimens. (**b**) Three-dimensional view of the specimens.

2.2. Fatigue Test Procedures

The servo-hydraulic testing machine MTS809 with the capability of 100 kN in axial load (Figure 2) was used to accomplish the fatigue test at room temperature. The constant amplitude fatigue experiments were conducted at a loading rate of 50 Hz with the stress ratios $R = 0.1$ ($R = \sigma_{min} / \sigma_{max}$) by using sinusoidal waveform control.

Figure 2. MTS (Systems Corporation) fatigue tensile testing machine.

In advance, a large number of tests had been conducted to determine the fatigue life of the specimens experimentally, ranging from 10,779 to 11,453 cycles at a loading frequency of 50 Hz and stress ratios $R = 0.1$ ($R = \sigma_{min}/\sigma_{max}$) by using sinusoidal waveform control. To reduce accidental experimental errors, the fatigue test was divided into three groups, and each group consisted of 5 cyclic loading stages (20,000 loading cycles, 40,000 loading cycles, 60,000 loading cycles, 80,000 loading cycles, 100,000 loading cycles), which were based on the maximum number of cycles determined as 100,000 cycles. These 5 stages were conducted on the corresponding 5 samples in each group, as shown in Figure 3.

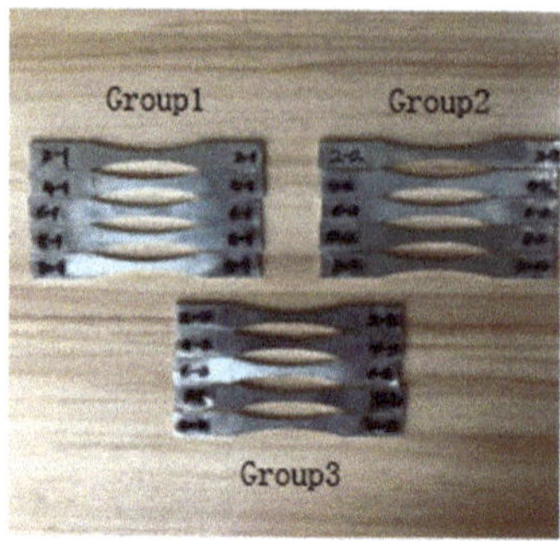

Figure 3. Test specimens.

3. Acquisition and Analysis of Defect Characteristics Information

3.1. Acquisition of Defect by X-ray Micro-Computed Tomography (MCT)

X-ray micro-computed tomography (MCT) (Vendor: Pheonix, AZ, USA), as a nondestructive technique, was used to get a quantitative estimate of mesoscopic voids in the damage specimens (Figure 4). In this process, an X-ray beam was focused on a particular region called the region of interest (ROI) of the specimen. ROI scan of the gauge length portion of the fatigue specimen could provide a deep insight into the shape and location distribution of voids, which is otherwise difficult to detect using any other conventional techniques, such as ultrasonic testing technology. A complete 360° scanning about the rotation axis of the specimen was carried out, and a total of 1600 images were taken. The exposure time for each image was 500 ms. The complete CT TIF section images of the ROI region were carried out by using Visual studio software from 2D data of each image. Four TIF diagrams selected randomly for specimen labeled 8-3 are shown in Figure 5; it would show different degrees of volume defects depending on perspectives.

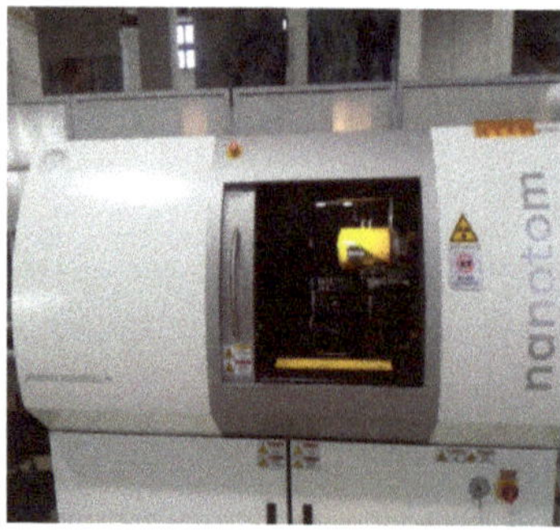

Figure 4. X-ray micro-computed tomography system.

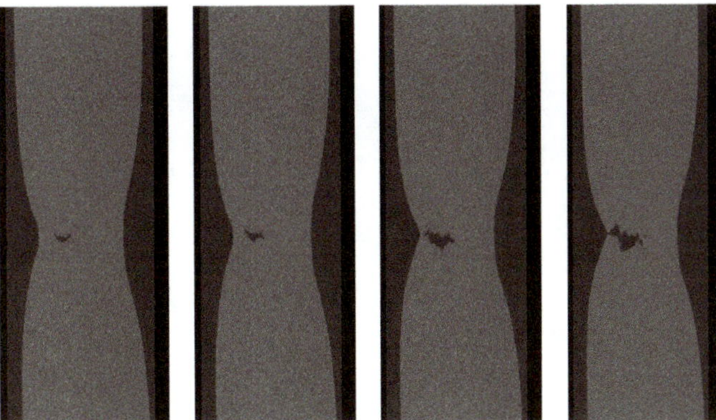

Figure 5. TIF slice diagram with different volume defects.

3.2. Visualization Processing of Defects Based on AVIZO

An effective visualization processing of slice data generated from CT was performed through importing TIF images with defect information into AVIZO, and the modular processing was conducted, as shown in Figure 6.

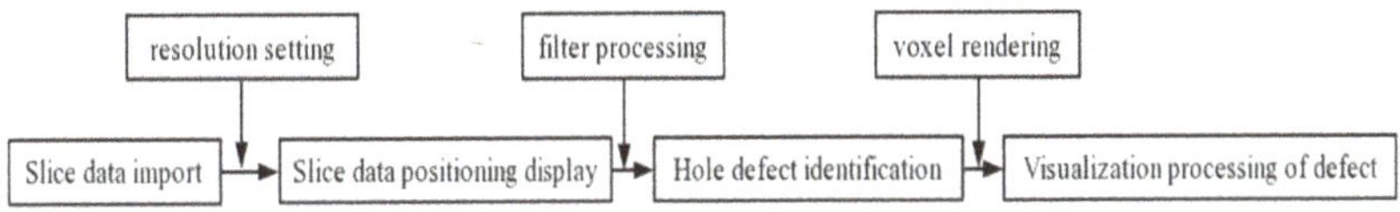

Figure 6. The flow chart of modular processing.

The resolution was set to 3 μm (CT scanning accuracy) in the process of importing TIF images into Avizo. The ortho-slice module was added to perform the TIF images data positioning explicit, as shown in Figure 7a. The non-local means filter module was used to filter the slice to remove the influence of the material itself on the hole defect recognition. The effect of TIF slice after noise removal is shown in Figure 7b. Comparing Figure 7a,b, it could be seen that the brightness of Figure 7b after noise removal was improved. The two-dimensional slice with noise removal was subjected to threshold segmentation through the interactive-threshold module to separate the matrix material from the void defect. The selection of the threshold was controlled during the segmentation process. If the threshold was too large, the matrix material would default to the void defect; and if the threshold was too small, the identified hole defect would be less than the actual number. Both of these situations were not conducive to subsequent defect analysis. The TIF diagram after segmentation and identification is shown in Figure 7c. The blue dot-like area in the figure indicates the void defect.

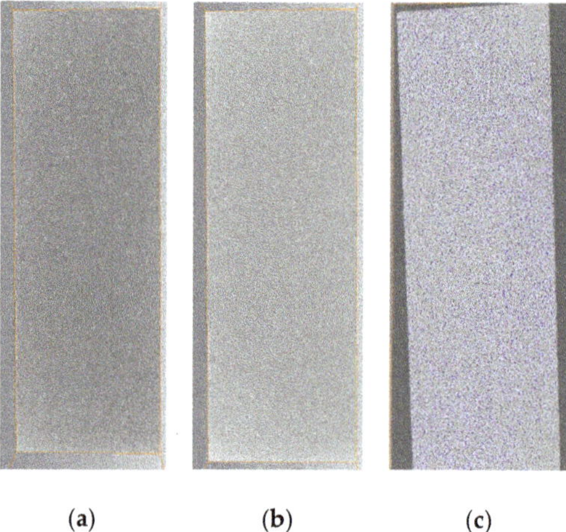

(a)   (b)   (c)

Figure 7. TIF diagram display. (**a**) TIF diagram after positioning display processing, (**b**) TIF diagram after noise filtering, (**c**) TIF diagram after threshold segmentation processing.

Moreover, voxel rendering was performed on the void defect by volume rendering, and the damage defect was three-dimensionally displayed and reconstructed to visually determine the position and degree of the damage. The 3D visualization of internal void defects for specimen labeled 8-3 is given in Figure 8.

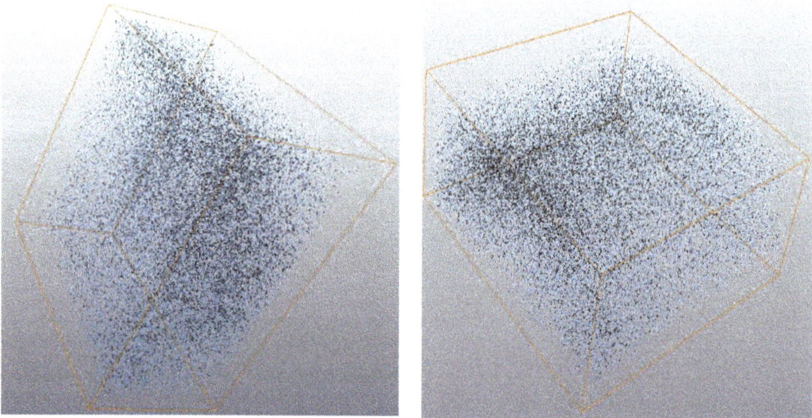

Figure 8. The 3D visualization of internal void defects for specimen labeled 8-3 from different perspectives.

3.3. Extraction and Analysis of Defect Characteristic Information

The characteristic information of defect (including void defect position, volume size, etc.) was obtained through the label analysis module for each cycle stage 3D model. Table 2 shows characteristic statistics of damage defects in different cycles.

Table 2. Characteristic statistics of damage defects in a different cycle.

Load Sequence ($\times 10^4$)	Stage	2	4	6	8	10
Total number of voids (n)	Group1	35,106	215,398	125,703	35,684	20,982
	Group2	36,061	233,821	104,895	30,569	16,456
	Group3	34,854	190,636	76,695	24,534	18,672
The volume of maximum void (μm^3)	Group1	972	1810	1970	3210	5671
	Group2	1190	2161	2275	3841	5319
	Group3	1021	2013	2144	3511	4782
Maximum damage surface area (μm^2)	Group1	0.0022	0.0392	0.0345	0.0596	0.24
	Group2	0.0151	0.0275	0.0226	0.0459	0.197
	Group3	0.0073	0.0157	0.0553	0.0412	0.211
Porosity (%)	Group1	0.0004	0.0046	0.0031	0.0287	0.2421
	Group2	0.0009	0.0025	0.0673	0.0562	0.3866
	Group3	0.0006	0.0051	0.0242	0.0901	0.5293

The variation of the defect characterization with the different loading stages is illustrated in Figure 9. Figure 9a shows that the number of voids increased first to the peak and then decreased rapidly. Figure 9b–d is plotted for the variation of the maximum void, maximum damage surface, and porosity with the number of cycles. It could be seen that the rate of damage growth was not obvious at the beginning loading stage; however, at the later stage of loading, the damage growth rate increased sharply with the growth and association of defects.

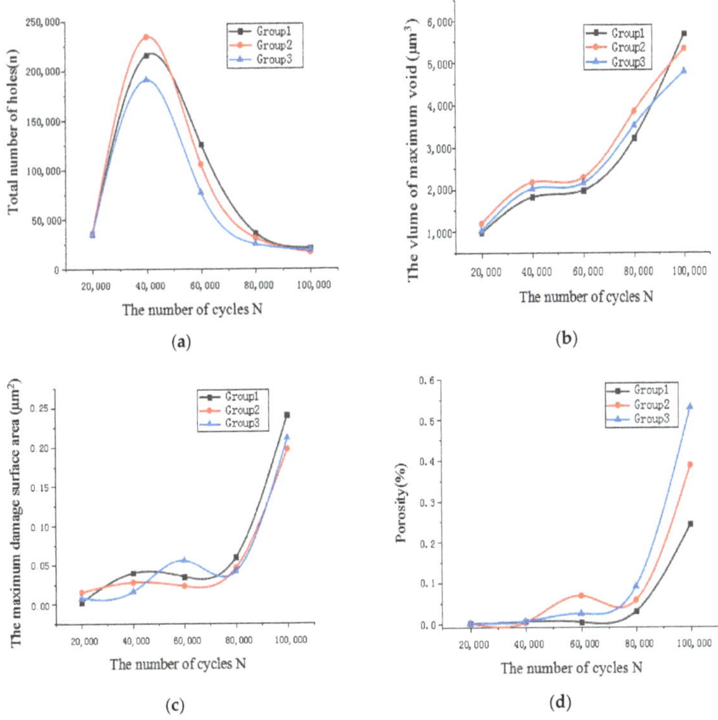

Figure 9. Variation of the voids damage characterization with the number of cycles N. (**a**) Variation of the total number of voids with the number of cycles N. (**b**) Variation of the volume of maximum void with the number of cycles N. (**c**) Variation of the maximum damage surface area with the number of cycles N. (**d**) Variation of the Porosity with the number of cycles N.

According to the variation of these damage characterizations (the number of hole defects, maximum hole, maximum damage surface, and porosity) with respect to the number of cycles, it could be concluded that the damage behavior mainly manifested that the initiation of void defects led to a slower growth of damage in the early loading stage. However, the growth rate of damage increased rapidly in the subsequent loading stages with the growth of micro-voids and the connection between adjacent voids.

The fatigue damage evolution process could be effectively characterized by the above damage parameters. Although some damage characterizations had volatility in the individual cycle stages, the overall tendency was consistent with the damage evolution law.

4. Damage Model

4.1. Establishment of an Equivalent Damage Model

It was difficult to conduct a damage evolution analysis and reduce the calculation time due to the shape and position of the plenty of mesoscopic defects inside the specimen. Therefore, a simplified equivalent multiscale damage model, which could reflect both mesoscopic damage and macroscopic fractures based on the grading of an internal defect, was established.

The equivalent mesoscopic void defects volume V_n and position information P_n were calculated based on original damage voids feature information. The volume of voids appeared in three orders of magnitude (10^{-6} mm^3, 10^{-7} mm^3, and 10^{-8} mm^3), which were, respectively, recorded as level 1, level 2, and level 3. The volume and position information of nth void for level 1 defects were recorded as V_{1n}, P_{1n}, and the remaining level 2 and level 3 were, respectively, recorded as V_{2m}, P_{2m} for m^{th} void and V_{3q}, P_{3q} for q^{th} void. The volume of the level 3 voids (10^{-8} mm^3) was averaged according to Equation (1) to obtain the average volume $\overline{V_3}$ since the volume of the level 3 voids was small. The level 1 voids were selected as base points, which were the main points with volume and location information; the level 2 and level 3 voids were selected as the reference points; the smallest volume in the level 3 voids was eliminated based on the effect of size on defect. Three levels (level 1, level 2, and level 3) of voids position information were placed in matrices P_1, P_2, and P_3, respectively. Matrix P_1 was a position matrix for the base points, and matrices P_2 and P_3 were recorded as the position matrices for the reference points, respectively, as shown in Equation (2).

$$\overline{V_3} = \frac{\sum\limits_{i=1}^{q} V_{3i}}{q} \tag{1}$$

$$P_1 = \begin{bmatrix} x_{11} & y_{11} & z_{11} \\ x_{12} & y_{12} & z_{12} \\ \vdots & \vdots & \vdots \\ x_{1n} & y_{1n} & z_{1n} \end{bmatrix} = \begin{bmatrix} p_{11} \\ p_{12} \\ \vdots \\ p_{1n} \end{bmatrix}, P_2 = \begin{bmatrix} x_{21} & y_{21} & z_{21} \\ x_{22} & y_{22} & z_{22} \\ \vdots & \vdots & \vdots \\ x_{2m} & y_{2m} & z_{2m} \end{bmatrix} = \begin{bmatrix} p_{21} \\ p_{22} \\ \vdots \\ p_{2m} \end{bmatrix}, P_3 = \begin{bmatrix} x_{31} & y_{31} & z_{31} \\ x_{32} & y_{32} & z_{32} \\ \vdots & \vdots & \vdots \\ x_{3q} & y_{3q} & z_{3q} \end{bmatrix} = \begin{bmatrix} p_{31} \\ p_{32} \\ \vdots \\ p_{3q} \end{bmatrix}, \tag{2}$$

$p_{1n} = \begin{bmatrix} x_{1n} & y_{1n} & z_{1n} \end{bmatrix}$ is a coordinate vector of the nth void of level 1, and $p_{2m} = \begin{bmatrix} x_{2m} & y_{2m} & z_{2m} \end{bmatrix}$ and $p_{3q} = \begin{bmatrix} x_{3q} & y_{3q} & z_{3q} \end{bmatrix}$ are the coordinate vector of the mth void of the level 2 and the qth void of the level 3, respectively. The MATLAB was applied to solve the aggregation problem through the distance d between the reference points position vectors p_{2m}, p_{3q} and the base points position vector p_{1n}.

$$
\begin{aligned}
d_1^{(1)} &= \left| p_{11} - p_{21} \right| = \sqrt{(x_{11} - x_{21})^2 + (y_{11} - y_{21})^2 + (z_{11} - z_{21})^2}, \\
d_2^{(1)} &= \left| p_{11} - p_{22} \right| = \sqrt{(x_{11} - x_{22})^2 + (y_{11} - y_{22})^2 + (z_{11} - z_{22})^2}, \\
&\qquad\qquad\qquad\vdots \\
d_m^{(1)} &= \left| p_{11} - p_{2m} \right| = \sqrt{(x_{11} - x_{2m})^2 + (y_{11} - y_{2m})^2 + (z_{11} - z_{2m})^2}, \\
d_{m+1}^{(1)} &= \left| p_{11} - p_{31} \right| = \sqrt{(x_{11} - x_{31})^2 + (y_{11} - y_{31})^2 + (z_{11} - z_{31})^2}, \\
&\qquad\qquad\qquad\vdots \\
d_{m+q}^{(1)} &= \left| p_{11} - p_{3q} \right| = \sqrt{(x_{11} - x_{3q})^2 + (y_{11} - y_{3q})^2 + (z_{11} - z_{3q})^2},
\end{aligned}
\tag{3}
$$

From Equation (3), the distance between the first base point p_{11} and the remaining reference points could be obtained, and the result obtained was recorded as a vector D_1. Similarly, the distance from the nth base point to the remaining reference points could be calculated and recorded as a vector D_n, as shown in Equation (4).

$$
D = \begin{bmatrix} D_1 \\ D_2 \\ \vdots \\ D_n \end{bmatrix} = \begin{bmatrix}
d_1^{(1)} & d_2^{(1)} & \cdots & d_{m+q}^{(1)} \\
d_1^{(2)} & d_2^{(2)} & \cdots & d_{m+q}^{(2)} \\
\vdots & \vdots & \cdots & \vdots \\
d_1^{(n)} & d_2^{(n)} & \cdots & d_{m+q}^{(n)}
\end{bmatrix}
\tag{4}
$$

While the distance $d_o^{(n)}$ ($1 \leq o \leq m + q$) in matrix D was less than 3 mm, which was determined according to the simulation analysis by Digmat software, the volume information of the reference points with $d_o^{(n)}$ was retained and recorded as $V_{2r}^{(n)}$, $\overline{V_3^{(n)}}$. Where $V_{2r}^{(n)}$ is the rth volume of level 2 void, and $\overline{V_3^{(n)}}$ is the average volume of level 3 voids, respectively. In addition, Table 3 shows that the distance d between each pair of the base points was larger than 3 mm, eliminating the interaction between the base points. The effective voids volume V_n and position information P_n in the equivalent model were derived from Equations (5) and (6).

$$
V_n = V_{1n} + \sum_{i=1}^{r} V_{2i}^{(n)} + w\overline{V_3^{(n)}} \quad (1 \leq r \leq m,\ 1 \leq w \leq q)
\tag{5}
$$

$$
P_n = P_{1n}
\tag{6}
$$

Table 3. Distance d between each pair of base points.

Reference Point (n)	1	2	3	4	$\cdots$	n
1	0	3.6457	5.1247	4.2487	$\cdots$	3.8546
2	3.6457	0	4.2136	4.8712	$\cdots$	5.0147
3	5.1247	4.2136	0	3.0014	$\cdots$	3.2387
4	4.2478	4.8712	3.0014	0	$\cdots$	3.5601
$\vdots$	$\vdots$	$\vdots$	$\vdots$	$\vdots$	$\cdots$	$\vdots$
n	3.8546	5.0147	3.2387	3.5601	$\cdots$	0

The schematic diagram of the distance between the base point voids and the reference point voids is shown in Figure 10.

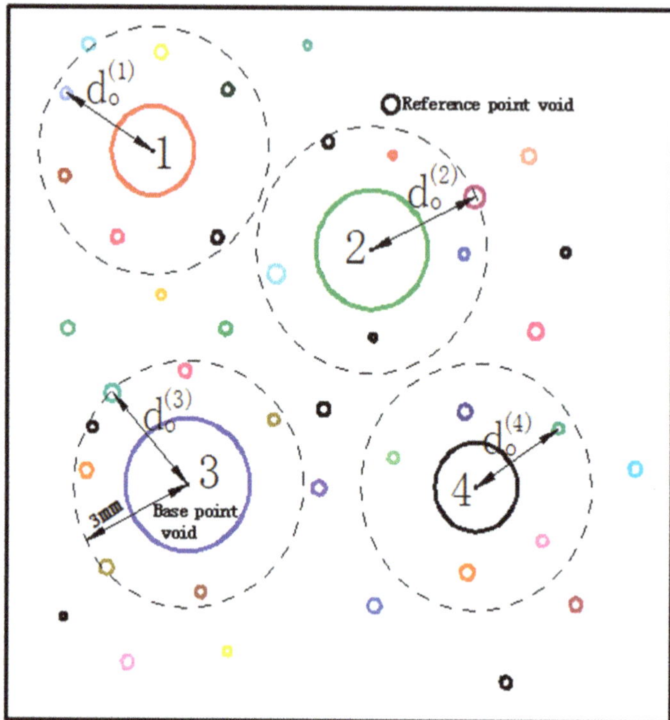

Figure 10. Schematic diagram of the distance between the base point voids and the reference point voids.

In addition, the voids defect was equivalent to a sphere model based on the simulation analysis by Digmat software. A lossless solid model was created by ABAQUS, and the equivalent position information P_n was assigned to the corresponding void defect model after the equivalent void model was created by the ABAQUS subroutine based on the equivalent void volume V_n. Then, the voids defect model assigned to the position information was moved to the corresponding position of the solid model by the ABAQUS subroutine and subjected to an ablation processing. Finally, the damage model with the equivalent void information was obtained. The schematic diagram of the equivalent damage model created by ABAQUS subroutine for specimen labeled 8-3 is demonstrated in Figure 11.

Damage model of the nth equivalent void

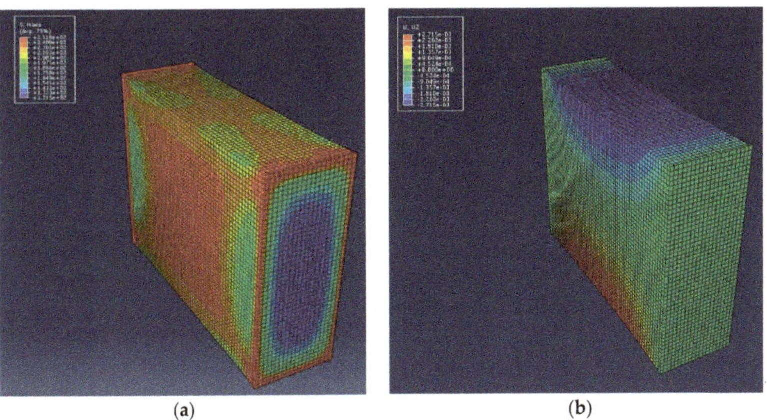

Figure 11. Schematic diagram of the equivalent damage model created by ABAQUS subroutine for specimen labeled 8-3.

4.2. Fatigue Model Analysis and Validation

The equivalent multiscale damage evolution model was subjected to finite element tensile analysis by using displacement control in ABAQUS. The simulation results of the multiscale damage model of a fatigue specimen labeled 8-3 is shown in Figure 12. The effective Young's modulus E_{ds} obtained by the finite element analysis of the equivalent multiscale damage evolution model of three groups of fatigue specimens was fitted into a curve, as shown by the dotted line in Figure 13.

(a)   (b)

Figure 12. The tensile simulation results of damage model of specimens labeled 8-3. (**a**) Stress cloud map, (**b**) strain cloud map.

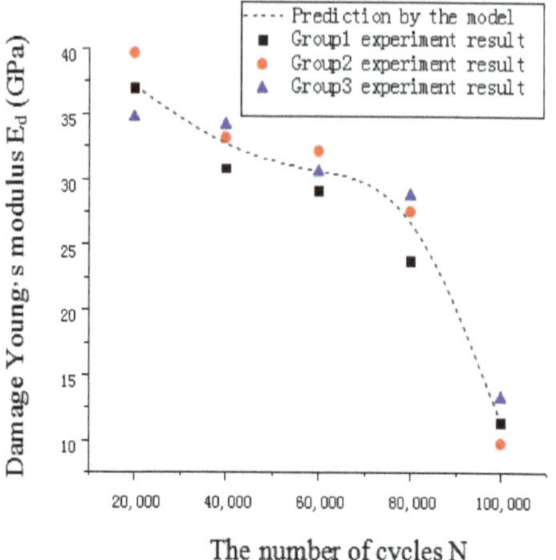

Figure 13. The verification of the fatigue damage evolution model.

In addition, the tensile experiments were conducted at a loading rate of 0.2 mm/min by using displacement control in electronic tensile testing machine to obtain damage Young's modulus E_{dt} of three groups of fatigue specimens. Table 4 shows the damage Young's modulus E_{dt} of three groups of fatigue specimens after the quasi-static tensile test.

Table 4. The Young's modulus E_{dt} of three groups of fatigue specimens after the tensile test.

load Sequence ($\times 10^4$)	Stage Group	2	4	6	8	10
Damage Young's modulus E_{dt} (GP$_a$)	Group1	36.581	31.816	29.109	24.169	11.356
	Group2	39.647	33.169	32.147	27.549	9.365
	Group3	34.946	34.149	30.564	28.764	13.248

In order to verify the proposed multiscale damage evolution model, the comparisons between the curve, predicted by the finite element numerical model parameters E_{ds}, obtained from ABAQUS and the tensile experimental results (Table 4) are shown in Figure 13. The comparisons showed that the fatigue damage evolution curves predicted by the multiscale damage evolution model agreed well with the experimental results, indicating that the proposed multiscale damage evolution model could describe not only macro-scale fatigue damage evolution process by using Young's modulus E but also describe the behavior of voids initiation and growth in mesoscopic scale. Therefore, the model built a bridge that used the effective Young's modulus E at the macro-scale to characterize the mesoscopic damage evolution.

5. Fatigue Life Prediction

Fatigue lives were predicted by establishing the relationship between the several important characteristic parameters: the damage Young's modulus E_d, the damage variable D, and the number of cycles.

By the continuous damage mechanism (CDM) [35,36] theory, the damage variable resulted from the initiation and growth of microvoids was often used to describe the degradation of material properties. For isotropic metal materials, the stiffness degradation represented by damage variable D is given by.

$$D = \frac{E - E_d}{E} \tag{7}$$

where E represents Young's modulus of material without damage, and E_d is the effective Young's modulus with damage.

The damage variable D of each loading stage was calculated by Equation (7) based on the effective Young's modulus E_d of each fatigue specimen after static tensile test in ABAQUS. Consequently, the relationship between damage parameter D and fatigue life in the fatigue process could be obtained. The variation of the damage variable D and the damage Young's modulus E_d with the number of loading cycles N is, respectively, plotted in Figure 14a,b.

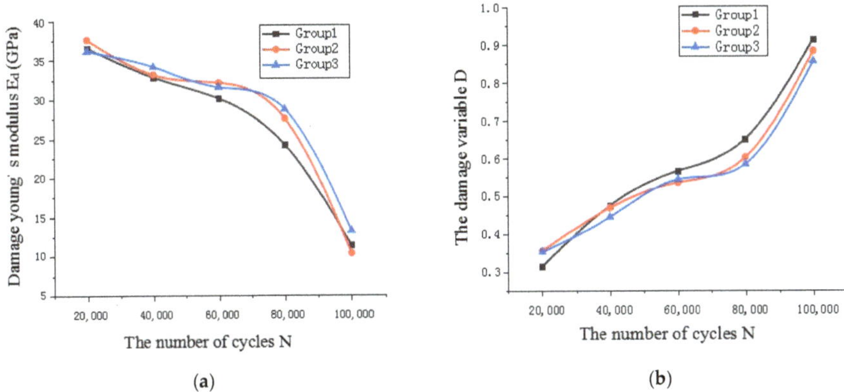

(a)   (b)

Figure 14. (a) Variation of Young's damage modulus E_d with the number of cycles N. (b) Variation of the damage variable D with the number of cycles N.

It could be seen from Figure 14a,b that the damage variable D increased with the increase of the number of loading cycles N, and the damage Young's modulus E_d decreased gradually. However, the damage Young's modulus E_d decreased drastically if the damage degree D exceeded 0.1, resulting in a sharp degradation of material properties. Moreover, Figure 14b shows that the damage parameter D of three groups of curves increased slowly in the early stage, which is the stable initiation stage of voids, and accounted for 10% to 80% of total life. After that, the damage parameter D increased rapidly in the unstable propagation stage of voids. However, no matter which damage stage the material was in, fatigue life could also be predicted conveniently based on the calculated damage variable D by the proposed multiscale damage evolution model for nondestructive detection.

6. Conclusions

The effectiveness of our multiscale damage evolution model was confirmed by using the experimental data of fatigue damage accumulation during the stages of voids initiation and growth. After that, the developed model was, respectively, applied to explain the effect of load sequence on fatigue life. In addition, the fatigue life of metal components and structures due to mesoscopic voids and growth and linkage was calculated from the damage variable D and the effective Young's modulus E_d at the macro-scale. According to the results of this study, the following conclusions could be presented:

(1) The proposed method of defect classification was effective to realize the three-dimensional reconstruction of the mesoscopic defects based on the mesoscopic defect data obtained by CT scanning technology.

(2) A new multiscale damage evolution model for fatigue damage accumulation had been developed, which built a bridge to describe the continuous average damage evolution process for metal fatigue components and structures in mesoscopic scale by macro damage variables for understanding metal fatigue failure mechanisms.

(3)   The fatigue life was predicted with the damage data measured by nondestructive testing technology (CT scanning technology, etc.) based on the effectiveness of the multiscale damage evolution model.

Author Contributions: Conceptualization, H.C. and J.S.; methodology, Y.Y.; software, Y.B.; validation, Y.B., Y.L., and S.Y.; formal analysis, Y.B.; investigation, S.Y.; resources, H.C.; data curation, Y.B. and Y.L.; writing—original draft preparation, Y.B.; writing—review and editing, Y.B., Y.Y., and J.S.; supervision, H.C.; funding acquisition, H.C.

Funding: This research was funded by the Natural Science Foundation of Shanghai (18ZR1416500) and Development Fund for Shanghai Talents, and the "Shuguang Program" supported by the Shanghai Education Development Foundation and Shanghai Municipal Education Commission.

Conflicts of Interest: The authors declare no conflict of interest.

References

1.   Kaynak, C.; Ankara, A.; Baker, T.J. Effects of short cracks on fatigue life calculations. *Int. J. Fatigue* **1996**, *18*, 25–31. [CrossRef]

2.   Lillbacka, R.; Johnson, E.; Ekh, M. A model for short crack propagation in polycrystalline materials. *Eng. Fract. Mech.* **2006**, *73*, 223–232. [CrossRef]

3.   Cheung, M.S.; Li, W.C. Probabilistic fatigue and fracture analyses of steel bridges. *Struct. Saf.* **2003**, *23*, 245–262. [CrossRef]

4.   Kachanov, L.M. Time of the rupture process under creep condition. *TVZ Akad. Nauk. Ssr Otd. Tech. Nauk.* **1958**, *8*, 26–31.

5.   Pathan, A.K.; Bond, J.; Gaskin, R.E. Sample preparation for scanning electron microscopy of plant surfaces—Horses for courses. *Micron* **2008**, *39*, 1049–1061. [CrossRef]

6.   Risitano, A.; Risitano, G. Cumulative damage evaluation of steel using infrared thermography. *Appl. Fract. Mech.* **2010**, *54*, 82–90. [CrossRef]

7.   Fan, J.L.; Guo, X.L.; Wu, C.W.; Zhao, Y. Research on fatigue behavior evaluation and fatigue fracture mechanisms of cruciform welded joints. *Mater. Sci. Eng.* **2011**, *A528*, 8417–8427. [CrossRef]

8.   Fan, J.L.; Guo, X.L.; Wu, C.W. A new application of the infrared thermography for fatigue evaluation and damage assessment. *Int. J. Fatigue* **2012**, *44*, 1–7. [CrossRef]

9.   Belaadi, A.; Bezazi, A.; Bourchak, M.; Scarpa, F. Tensile static and fatigue behaviour of sisal fibres. *Mater. Des.* **2013**, *46*, 76–83. [CrossRef]

10.   Zhou, Y.F.; Jerrams, S.; Chen, L. Multi-axial fatigue in magnetorheological elastomers using bubble inflation. *Mater. Des.* **2013**, *50*, 68–71. [CrossRef]

11.   Azadi, M.; Farrahi, G.H.; Winter, G.; Eichlseder, W. Fatigue lifetime of AZ91 magnesium alloy subjected to cyclic thermal and mechanical loadings. *Mater. Des.* **2014**, *53*, 639–644. [CrossRef]

12.   Moreno-Navarro, F.; Rubio-Gámez, M.C. Mean damage parameter for the characterization of fatigue cracking behavior in bituminous mixes. *Mater. Des.* **2014**, *54*, 748–754. [CrossRef]

13.   Li, D.Z.; Han, L.; Thornton, M.; Shergold, M.; Williams, G. The influence of fatigue on the stiffness and remaining static strength of self-piercing riveted aluminium joints. *Mater. Des.* **2014**, *54*, 301–314. [CrossRef]

14.   Shiri, S.; Pourgol-Mohammad, M.; Yazdani, M. Probabilistic Assessment of Fatigue Life in Fiber Reinforced Composites. In Proceedings of the ASME 2014 International Mechanical Engineering Congress and Exposition, Montreal, QC, Canada, 10–14 November 2014; p. V014T08A018.

15.   Shiri, S.; Pourgol-Mohammad, M.; Yazdani, M. Effect of strength dispersion on fatigue life prediction of composites under two-stage loading. *Mater. Des.* **2015**, *65*, 1189–1195. [CrossRef]

16.   Shiri, S.; Yazdani, M.; Pourgol-Mohammad, M. A fatigue damage accumulation model based on stiffness degradation of composite materials. *Mater. Des.* **2015**, *88*, 1290–1295. [CrossRef]

17.   Miner, M.A. Cumulative damage in fatigue. *J. Appl. Mech.* **1945**, *67*, 159–164.

18.   Marco, S.M.; Starkey, W.L. A concept of fatigue damage. *Trans. ASME* **1954**, *76*, 627–632.

19.   Henry, D.L. A theory of fatigue damage accumulation in steel. *Trans. ASME* **1955**, *77*, 913–918.

20.   Gatts, R.R. Application of a cumulative damage concept to fatigue. *J. Basic Eng. Trans. ASME* **1961**, *83*, 529–540. [CrossRef]

21. Manson, S.S.; Freche, J.C.; Ensign, C.R. *Application of a Double Linear Damage Rule to Cumulative Fatigue*; NASA TN D-3839; NASA: Washington, DC, USA, 1967.
22. Chaboche, J.L.; Lesne, P.M. A non-linear continuous fatigue damage model. *Fatigue Fract. Eng. Mater. Struct.* **1988**, *11*, 1–17. [CrossRef]
23. Manson, S.S.; Halford, G.R. Practical implementation of the double linear damage rule and damage curve approach for treating cumulative fatigue damage. *Int. J. Fract.* **1981**, *17*, 169–192. [CrossRef]
24. Franke, L.; Dierkes, G. A non-linear fatigue damage rule with an exponent based on a crack growth boundary condition. *Int. J. Fatigue* **1999**, *21*, 761–767. [CrossRef]
25. Miller, K.J.; Mohanmed, H.J.; DelosRios, E.R. *The Behaviour of Short Fatigue Cracks*; Mechanical Engineering Publications: London, UK, 1986.
26. Angelova, D.; Akid, R. A note on modelling short fatigue crack behaviour. *Fatigue Fract. Eng. Mater. Struct.* **1998**, *21*, 771–779. [CrossRef]
27. Polák, J.; Liškutíne, P. Nucleation and short crack growth in fatigued polycrystalline copper. *Fatigue Fract. Eng. Mater. Struct.* **1990**, *13*, 119–133. [CrossRef]
28. McClintock, F.A. A criterion for ductile fracture by the growth of holes. *J. Appl. Mech. Trans. Asme* **1968**, *35*, 363–371. [CrossRef]
29. Rice, J.R.; Tracey, D.M. On the ductile enlargement of voids in triaxial stress fifields. *J. Mech. Phys. Solids* **1969**, *17*, 201–217. [CrossRef]
30. McDowell, D.L.; Gall, K.; Horstemeyer, M.F.; Fan, J. Microstructure-Based fatigue modeling of cast a356-t6 alloy. *Eng. Fract. Mech.* **2003**, *70*, 49–80. [CrossRef]
31. Leuders, S.; Thöne, M. On the mechanical behaviour of titanium alloy TiAl6V4 manufactured byselective laser melting: Fatigue resistance and crack growth performance. *Int. J. Fatigue* **2013**, *48*, 300–307. [CrossRef]
32. Hu, Y.N.; Wu, S.C.; Song, Z.; Fu, Y.N.; Yuan, Q.X.; Zhang, L.L. Effect of microstructural features on the failure behavior of hybrid laser welded AA7020. *Fatigue Fract. Eng. Mater.* **2018**, *41*, 2010–2023. [CrossRef]
33. Desmorat, R.; Tertre, A.D.; Gaborit, P. Multiaxial haigh diagrams from incremental two scale damage analysis. *AerospaceLab* **2015**, 1–15.
34. Wan, H.L.; Wang, Q.; Chenxue, J.; Zhang, Z. Multi-Scale damage mechanics method for fatigue life prediction of additive manufacture structures of Ti-6Al-4V. *Mater. Sci. Eng. A* **2016**, *669*, 269–278. [CrossRef]
35. Lemaitre, J.; Chaboche, J.L. *Mechanics of Solid Materials*; Cambridge University Press: Cambridge, UK, 1994.
36. Lemaitre, J.; Desmorat, R. *Engineering Damage Mechanics: Ductile, Creep, Fatigue and Brittle Failures*; Springer Science & Business Media: Cachan, France, 2005.

 © 2019 by the authors. Licensee MDPI, Basel, Switzerland. This article is an open access article distributed under the terms and conditions of the Creative Commons Attribution (CC BY) license (http://creativecommons.org/licenses/by/4.0/).

Article

Monitoring of Fatigue Crack Propagation by Damage Index of Ultrasonic Guided Waves Calculated by Various Acoustic Features

Hashen Jin, Jiajia Yan, Weibin Li * and Xinlin Qing *

School of Aerospace Engineering, Xiamen University, Xiamen 361005, China;
35120170154978@stu.xmu.edu.cn (H.J.); yanjj210@163.com (J.Y.)
* Correspondence: liweibin@xmu.edu.cn (W.L.); xinlinqing@xmu.edu.cn (X.Q.);
  Tel.: +86-187-5021-8190 (W.L.); +86-187-5928-7299 (X.Q.)

Received: 18 September 2019; Accepted: 3 October 2019; Published: 11 October 2019

Abstract: Under cyclic and repetitive loads, fatigue cracks can be further propagated to a crucial level by accumulation, causing detrimental effects to structural integrity and potentially resulting in catastrophic consequences. Therefore, there is a demand to develop a reliable technique to monitor fatigue cracks quantitatively at an early stage. The objective of this paper is to characterize the propagation of fatigue cracks using the damage index (DI) calculated by various acoustic features of ultrasonic guided waves. A hybrid DI scheme for monitoring fatigue crack propagation is proposed using the linear fusion of damage indices (DIs) and differential fusion of DIs. An experiment is conducted on an SMA490BW steel plate-like structure to verify the proposed hybrid DIs scheme. The experimental results show that the hybrid DIs from various acoustic features can be used to quantitatively characterize the propagation of fatigue cracks, respectively. It is found that the fused DIs calculated by the acoustic features in the frequency domain have an improved reliable manner over those of the time domain. It is also clear that the linear and differential amplitude fusion DIs in the frequency domain are more promising to indicate the propagation of fatigue cracks quantitatively than other fused ones.

Keywords: monitoring of fatigue crack; damage index; ultrasonic guided waves; sensor network; structural health monitoring

---

1. Introduction

Fatigue cracks, which originate from a damaged precursor at an imperceptible level under repetitive loading, is one of the cardinal reasons for the failure of metallic structures. It is reported that up to 90% of the failures of in-service metallic structures are typically caused by fatigue cracks [1], and the formation of an initial fatigue crack does not necessarily result in immediate failure for the real-world structure [2]. Under cyclic loads, fatigue cracks at the scale of a few micrometers are then accumulated into microcracks by accumulation, which can deteriorate continuously and eventually amalgamate to form macrocracks [3,4]. Under repetitive loads, the macrocracks can be further propagated to a crucial level at an amazing rate without sufficient warning, causing detrimental effects on structural integrity and potentially resulting in catastrophic consequences [5]. Therefore, the early perception of small-scale fatigue cracks has become a critical measure to ensure the durability, reliability, and integrity of engineering structures [6,7].

The majority of current guided wave-based non-destructive testing (NDT) and structural health monitoring (SHM) techniques [3,8–13] have been proposed for monitoring crack propagation in different engineering structures, which exploit the variations from temporal features associated with baseline signals. More research studies of temporal signal features primarily focus on the magnitude-based and energy-based signals [14,15], wave reflections or transmissions [16,17], energy dissipation [18], and mode conversions [19]. Meanwhile, acoustic emission (AE) technology is widely used for continuously monitoring fatigue cracks, and AE-based fatigue crack evaluation techniques are exploited based on counting the number of signals allured by crack propagation [20,21]. The scattering and attenuation of ultrasonic guided waves caused by fatigue cracks are applied to quantify the cracks' growth [22–24]. Although most approaches of temporal features have the capacity to locate fatigue cracks, characterizing the orientation and size of fatigue cracks accurately still remains an extremely challenging task [25], because small-scale fatigue cracks can lead to a feeble signal difference for the damage-scattered waves in the time domain. Therefore, signal features processing and analysis associated with the frequency domain are then extracted to characterize the fatigue crack propagation, because the signal features in the frequency domain are more stable, accurate, and recognizable than those of the temporal domain.

The ultrasonic guided waves methods based on damage indices (DIs) have become research hotspots in the field of damage detection over the years, which aim to highlight the variations of structural fatigue cracks. The accuracy and effectiveness of the techniques substantially rely on the means of the defined DI at each path [25], thus, there is an increasing interest in introducing DI for guided wave-based SHM [26]. Based on linear (time of flight (TOF) and energy) and nonlinear temporal features of guided waves, damage indices (DIs) are respectively constructed, which are used to locate fatigue damage near a rivet hole of an aluminum plate [3,9,25]. Moreover, the synthetic DI methods are easily interpretable, which can intuitively reflect the overall health status in the inspected structure. A hybrid DI characterization scheme for ultrasonic guided waves is developed and experimentally validated in aluminum plate specimens, which can be used to evaluate the statistical distribution of fatigue cracks under uncertainties and update the prognosis results. However, according to the aforementioned literature survey, damage indices are used to locate cracks and identify the existence of multi-scale cracks in practice [25]. The crack growth is a gradually patulous process, the ultrasonic guided waves DI approach has been rarely employed to characterize the crack propagation quantitatively. In addition, a hybrid DI scheme is proposed that is associated with various acoustic features, which can be used to indicate the fatigue crack propagation quantitatively by experimental validation. This is deemed as an improvement to characterize crack growth in current monitoring studies. Although it is a highly challenging assignment in perceiving small-scale fatigue cracks, it is also of vital importance for continuously monitoring the crack propagation [10].

The objective of this paper is to propose a technique to monitor the fatigue crack propagation in an SMA490BW steel plate-like structure. Ultrasonic guided waves DIs calculated by various acoustic features are used to track the crack propagation. The features of crack characterization are firstly expressed by magnitude/amplitude-based and energy-based DIs extracted from time-frequency domain signals. Based on the DI weights of each actuator–sensor path, a hybrid fusion scheme is then proposed to fuse all the available DIs of each path. Subsequently, an experimental setup of an active sensor network with a sparse transducer configuration is used for actuating and sensing guided waves. Finally, the different synthetic fusion DI algorithms of the S0 guided waves mode are calculated to emphatically characterize the crack propagation and destruction level of an in-service metallic structure.

2. Fundamental Theory

2.1. Propagation of Ultrasonic Guided Waves in SMA490BW Steel Plate

Figure 1 illustrates the dispersion curves of ultrasonic guided waves in an SMA490BW steel plate, which is calculated using DISPERSER®. Generally, the phase velocity and group velocity of the propagating guided waves are respectively described by C_p and C_g. The phase velocity is the speed of a particle moving at a specific frequency. The group velocity is the disseminated speed of all wave packets [27]. The phase velocity C_p has the relationship with the wavelength $\lambda_{guided\ wave}$ and frequency f, as shown in Equation (1).

$$C_p = f\lambda_{guided\ wave} \tag{1}$$

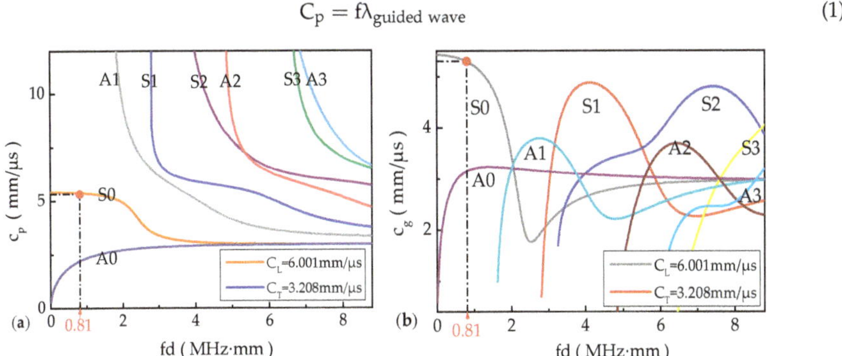

Figure 1. Dispersion curves of ultrasonic guided waves in an SMA490BW steel plate (**a**) Phase velocity curves; (**b**) Group velocity curves.

Since the selected excitation frequency of guided waves is 270 kHz, the uniform thickness of the SMA490BW steel plate is 3 mm, and the product of frequency–thickness is 0.81 MHz·mm, as marked in Figure 1. The S0 mode propagates at the fastest velocity among all available modes, thus, it can be completely separated from the other modes, boundary reflection, attenuation, and mode conversion signals. Therefore, the acquired signals of the S0 mode are adopted for the corresponding signal processing and analysis in this paper.

2.2. Features for Fatigue Crack Characterization

The TOF, magnitude/amplitude, and wave energy extracted from the captured signals are three sorts of the most representative ultrasonic guided waves features, which have proven effectiveness in locating gross damage [9,25,28]. In this study, the magnitude/amplitude and wave energy signals acquired from a piezoelectric active sensor network with a pitch–catch configuration are respectively extracted at different crack lengths for establishing magnitude/amplitude-based and energy-based DIs, which are opted to indicate guided waves signal features.

2.2.1. Magnitude/Amplitude-Based DIs

In this study, in order to interpret ultrasonic guided waves features acquired by individual actuator–sensor paths in an active sensor network. A damage index (DI) is established across the entire inspection area, within which fatigue cracks, if there are any, are expected to be visualized by an intuitive DI. Here, the DI is defined with the magnitude/amplitude alteration of a scattering wave caused by the occurrence of a crack at a particular actuator–sensor path in the inspection area, denoting a magnitude/amplitude-based DI.

$$DI(i,j)_{magnitude\text{-}k\text{-}t} = \frac{u(i,j)_{B\text{-}k}(t) - u(i,j)_{D\text{-}k}(t)}{u(i,j)_{B\text{-}k}(t)} \tag{2}$$

$$DI(i,j)_{amplitude\text{-}k\text{-}f} = \frac{u(i,j)_{B\text{-}k}(f) - u(i,j)_{D\text{-}k}(f)}{u(i,j)_{B\text{-}k}(f)} \tag{3}$$

In Equations (2) and (3), DI(i,j) stands for the DI calculated at the A_i–S_j actuator-sensor path, magnitude/amplitude refers to the temporal magnitude/spectrum amplitude features of an ultrasonic guided wave, and subscript k represents the kth (A_i–S_j) actuator–sensor path. The t and f refer to the time and frequency domain, respectively. $u(i,j)_{B\text{-}k}(t)$ and $u(i,j)_{D\text{-}k}(t)$ exclusively correspond to the baseline signal and current signal of the magnitude peaks of the S0 mode, which are acquired by the A_i–S_j actuator–sensor path in the time domain. Similarly, their corresponding benchmark signal and current signal in the frequency domain are $u(i,j)_{B\text{-}k}(f)$ and $u(i,j)_{D\text{-}k}(f)$. $u(i,j)_{B\text{-}k}(t) - u(i,j)_{D\text{-}k}(t)$ and $u(i,j)_{B\text{-}k}(f) - u(i,j)_{D\text{-}k}(f)$ are the scatter magnitude/amplitude signal of the S0 mode caused by fatigue cracks in the time and frequency domain. On account of Equations (2) and (3), the defined magnitude/amplitude-based DIs are used for quantitatively characterizing the crack propagation at each path in the inspected engineering structure.

2.2.2. Energy-Based DIs

The discrepancy in fatigue cracks (different distances to an actuator–sensor path, severities, and orientations [25]) can result in a distinct damage-scattered magnitude of ultrasonic guided waves. Therefore, the deviations of wave energy, which depend on the ultrasonic guided waves signals captured from the target structure (called damage or current signal) and its healthy counterpart acquired from a pristine benchmark (denoted baseline signal), can be employed to develop a DI. In consideration of the correlations between the current and baseline signals, as follows:

$$D(i,j)_{energy\text{-}k\text{-}t} = \frac{\int_{t1}^{t2} \left[u(i,j)_{B\text{-}k}(t) - u(i,j)_{D\text{-}k}(t) \right]^2 dt}{\int_{t1}^{t2} u(i,j)_{B\text{-}k}^2(t)dt} \tag{4}$$

$$D(i,j)_{energy\text{-}k\text{-}f} = \frac{\int_{f1}^{f2} \left[u(i,j)_{B\text{-}k}(f) - u(i,j)_{D\text{-}k}(f) \right]^2 df}{\int_{f1}^{f2} u(i,j)_{B\text{-}k}^2(f)df} \tag{5}$$

Equations (4) and (5) are similar to Equations (2) and (3), in which $DI(i,j)_{energy}$ alludes to the DI that is defined at the A_i–S_j path for ultrasonic guided waves signal features associated with energy. The baseline signal $u(i,j)_{B\text{-}k}(t)$ and damage signal $u(i,j)_{D\text{-}k}(t)$ in the time domain are acquired by the A_i–S_j actuator-sensor path, and their corresponding benchmark signal and current signal in the frequency domain severally are $u(i,j)_{B\text{-}k}(f)$ and $u(i,j)_{D\text{-}k}(f)$. $\int_{t1}^{t2} \left[u(i,j)_{B\text{-}k}(t) \right]^2 dt$ is the baseline signal energy of the S0 mode in the time domain. The t1 is the initial time position of the S0 mode, and the t2 is the end time position of the S0 mode. $\int_{f1}^{f2} \left[u(i,j)_{B\text{-}k}(f) \right]^2 df$ is the baseline energy signal of the S0 mode in the frequency domain. The f1 is the initial frequency position of the S0 mode, and the f2 is the end frequency position of the S0 mode. Similarly, $\int_{t1}^{t2} \left[u(i,j)_{B\text{-}k}(t) - u(i,j)_{D\text{-}k}(t) \right]^2 dt$ and $\int_{f1}^{f2} \left[u(i,j)_{B\text{-}k}(f) - u(i,j)_{D\text{-}k}(f) \right]^2 df$ are the scatter energy of the S0 mode caused by fatigue cracks in the time and frequency domain. An energy-based DI related with the guided waves signal feature can be constructed at each path, which is applied to quantify the crack growth and characterize the destruction level in the inspected structure range. Furthermore, the greater $DI(i,j)_{energy\text{-}k}$ obtained along with the A_i–S_j path, the larger the indicated fatigue crack propagation.

3. Fusion of DIs

Driven by the incentive to visualize the procedure of fatigue crack propagation, a damage index synthetic fusion algorithm of each actuator–sensor path is further developed in this study. Using a hybrid fusion method does not weaken the characterization of crack propagation, but it does intuitively and rapidly comprehend the overall health status within the inspection of the SMA490BW plate-like structure. The fatigue crack propagation can be described by the fitting correlation of hybrid fused DIs of various acoustic characteristics.

In terms of Equations (2)–(5), the extracted various signal features are used to establish different DIs. The corresponding magnitude/amplitude and energy DIs are calculated at each actuator–sensor path, which are applied to define the original DIs. The original DI has a prior perception for fatigue cracks on the current path. In practice, an original DI not only contains information associated with cracks, but also unwanted signal features such as ambient noise and measurement uncertainties, multiple guided waves modes conversion, and reflections from structural boundaries. These unwanted signal features possibly dilute crack-associated features and weaken the proportion from a single actuator–sensor path [25]. Therefore, based on the DI weights of different paths, the signal representations of the fused DIs are embodied by Equations (6)–(9). The ultimate fitting result is a straight line, which can be used to indicate the fatigue crack propagation effectively.

To summarize the above, each actuator–sensor path (N in total) offers an original DI via a pitch–catch configuration. Then, a hybrid fusion scheme is proposed to coalesce all the available DIs of each path in the sensor network, as follows:

$$\mathrm{DI}(i,j)_{\text{magnitude-fusion-t}} = \sum_{k=1}^{N}\left(\alpha_{\text{magnitude-k-t}}\mathrm{DI}(i,j)_{\text{magnitude-k-t}}\right) \tag{6}$$

$$\mathrm{DI}(i,j)_{\text{amplitude-fusion-f}} = \sum_{k=1}^{N}\left(\alpha_{\text{amplitude-k-f}}\mathrm{DI}(i,j)_{\text{amplitude-k-f}}\right) \tag{7}$$

$$\mathrm{DI}(i,j)_{\text{energy-fusion-t}} = \sum_{k=1}^{N}\left(\alpha_{\text{energy-k-t}}\mathrm{DI}(i,j)_{\text{energy-k-t}}\right) \tag{8}$$

$$\mathrm{DI}(i,j)_{\text{energy-fusion-f}} = \sum_{k=1}^{N}\left(\alpha_{\text{energy-k-f}}\mathrm{DI}(i,j)_{\text{energy-k-f}}\right) \tag{9}$$

where $\mathrm{DI}(i,j)_{\text{magnitude-fusion-t}}$ and $\mathrm{DI}(i,j)_{\text{energy-fusion-t}}$ are the ultimate fusion DIs in the time domain, meanwhile, $\mathrm{DI}(i,j)_{\text{amplitude-fusion-f}}$ and $\mathrm{DI}(i,j)_{\text{energy-fusion-f}}$ are the synthetic DIs in the frequency domain. The k stands for the kth actuator–sensor path, while α refers to the weights of each path. The incentive to build such a hybrid fusion scheme is twofold:

(i) An arithmetic fusion equally takes into account prior perceptions from each actuator–sensor path and well decentralizes individual contributions. Although the arithmetic fusion fully encompasses previous perceptions from all paths, information including ambient noise and measurement uncertainties are also engaged, which might overstate the presence of fatigue cracks and lead to false alarms, and

(ii) The synthesized fusion DI method is obtained by implementing a probability weighting operation on the original DIs of each actuator–sensor path. A hybrid DI scheme is constructed to enhance the approach tolerance for environmental noise, measurement uncertainties, and erroneous perceptions from individual paths.

The hybrid fusion scheme can be used to represent the fatigue crack propagation quantitatively, which effectively strengthens the individual original DIs of crack-related information and enhances the availability, stability, and signal-to-noise ratio of the fitting signals.

4. Experiment

4.1. Specimens

The research objective is to monitor the fatigue crack propagation by means of the DIs in ultrasonic guided waves. Furthermore, the material type of the bogie frame is SMA490BW steel. The intensity and rigidity of SMA490BW steel are confirmed in accordance with JISE4208 standard (railway vehicle bogie static load test method).

A 190 mm × 187 mm × 3 mm bogie steel plate is used in the experiment, whose material properties and chemical composition are shown in Tables 1 and 2. Using an active sensor network with a comparatively sparse transducer configuration, six manufactured identical PZT-5As are surface-mounted on the specimen by conductive epoxy. Each PZT-5A has a diameter of 8 mm and thickness of 0.48 mm. A_1–A_3 of PZT-5As are respectively used for actuators as ultrasonic guided waves inputs, whereas S_4–S_6 of PZT-5As are used for sensors, forming nine actuator–sensor paths: A_1–S_4, A_1–S_5, A_1–S_6, A_2–S_4, A_2–S_5, A_2–S_6, A_3–S_4, A_3–S_5, and A_3–S_6 are respectively named the first to ninth actuator–sensor paths, as schematically shown in Figure 2. Those paths are deployed by the physically existing actuators (A_1–A_3) and sensors (S_4–S_6), the actuator–sensor path network can also be formed to locate any position of interest within the inspection area virtually [25]. Fatigue cracks propagate approximately from 0.0 mm to 180.0 mm in steps of 5.0 mm, which are cut using a 1.0 mm thick saw blade. The visible fatigue crack through the plate thickness and in parallel with the 187.0 mm side is generated.

Table 1. Material parameters of the SMA490BW steel plate.

Material	D (mm)	ρ(kg/m^3)	f (kHz)	E (Gpa)	G (Gpa)	υ
SMA490BW	3	7850	270	210	80.7	0.3

Table 2. Chemical composition of the SMA490BW steel plate.

Material	C	Si	Mn	P	S	Cr	Ni	Cu	Other
SMA490BW	0.18	0.65	1.4	0.035	0.75	0.75	0.30	0.5	0.15

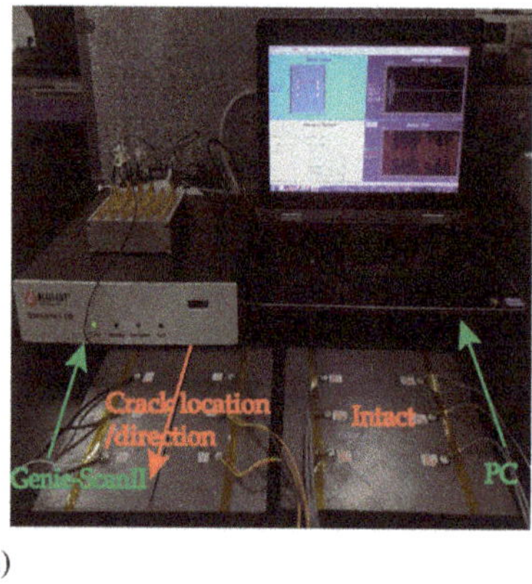

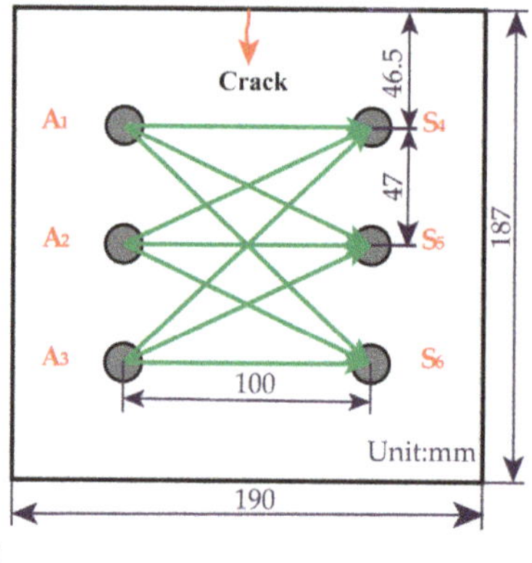

(a)   (b)

Figure 2. Experimental setup of an SMA490BW steel plate (a) The experimental equipment and specimen; (b) Schematic diagram of the specimen.

4.2. Experimental Setup

The 64-channel Scan-Genie II data acquisition integration system developed by Acellent Technologies, Inc. is used for signals excitation and reception. At each crack length measurement, the relevant ultrasonic guided wave signal responses are recorded. Considering that the signal energy of the guided waves is mostly concentrated in the vicinity of the narrowband probing frequency domain and the dispersion characteristics can also be reduced accordingly, the narrowband sinusoidal tone burst modulated by a Hanning window is calculated by Equation (10) as an excitation frequency. The excitation signal is applied to A_1–A_3 of PZT-5As, respectively, and the peak-to-peak voltage of the output signal is 50 V, it is amplified by the gain parameter setting before it is sent to A_1–A_3. The remaining S_4–S_6 of PZT-5As are used to measure the guided waves signals. In the meantime, the sampling rate is 24 MHz, the sampling point is 6000 and the quality of measurement is improved by averaging the signals with multiple acquisitions. Therefore, the output signal responses are obtained six times and averaged to improve the signal-to-noise ratio.

$$h(n) = \frac{1}{2}\left[1 - \cos\left(\frac{2\pi n}{N-1}\right)\right] (n = 0, 1, \dots, N-1) \tag{10}$$

4.3. Exciting Frequency Selection

Firstly, a mode-tuning experiment [7] is carried out to select the optimal excitation frequency that could maximize the magnitude/amplitude responses of the guided waves. The sweep ranges of central frequency are determined by considering the local resonance characteristics of PZT-5A. As a result, the driving frequencies are swept from 110 kHz to 330 kHz in steps of 20 kHz. Then, it is applied to one of the actuators to generate guided waves and the rest of the sensors are used for receiving. At each excitation frequency, the magnitude signals of ultrasonic guided waves are recorded. The local resonance frequency of PZT-5A measured by a WK-6500B impedance instrument is 265 kHz, as shown in Figure 3, and the resonance frequency [29] calculated by theoretical Equation (11) is 304 kHz. The central frequency error between theoretical calculation and experimental measurement is less than 13%, and the difference can be negligible. The frequency near the resonance frequency of PZT-5A is selected as the central excitation frequency of the experimental because the signals acquired at this frequency have a high stability and a high signal-to-noise ratio. Therefore, the excitation frequency of 270 kHz is chosen to excite guided waves in the rest of the studies. Figure 4 shows the excitation signal of five-cycle Hanning-windowed sinusoidal tone bursts at 270 kHz, it can also be seen that the amplitude of the central frequency from the spectrogram is maximum at 270 kHz.

$$f_{\text{res-PZT}} = \frac{\eta_n}{2\pi r} \sqrt{\frac{1}{\rho E(1 - v^2)}} \tag{11}$$

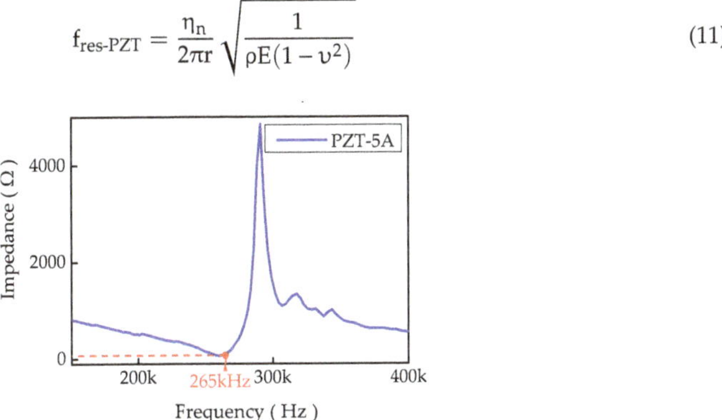

Figure 3. The measured impedance of PZT-5A by a WK-6500B impedance instrument.

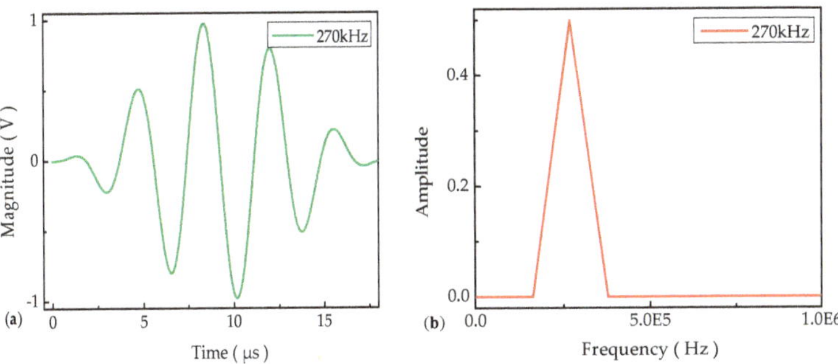

Figure 4. The 270 kHz excitation signals of five-cycle sinusoidal tone bursts modulated by Hanning-windowed (**a**) Excitation signal; (**b**) Excitation spectrum signal.

$\eta_n = \eta_n(v)$, in which n generally equals 1, $v_{\text{PZT}} = 0.35$, $\eta_1 = 2.0795$, r refers to the radius of PZT-5A, and E stands for the elastic modulus.

5. Results and Discussions

The measured guided wave signals from the damaged specimen and baseline signals in intact specimen are used to construct magnitude/amplitude-based and energy-based DIs for quantitatively characterizing crack growth in an SMA490BW steel plate-like structure along with the actuator–sensor paths. In cases where a crack is exactly located near the path or the crack size is equivalent to the same order of magnitude of the detection wavelength, the representative acoustic features of the measured guided wave are changed significantly, which corresponds to the larger DI coefficient between two signals. In contrast, the signal deviation would be trivial if the crack is far away from the actuator–sensor path or the crack size is much smaller than the probing wavelength [25]. In particular, the magnitude/amplitude-based and energy-based DIs have proven the susceptibility to signal magnitude/amplitude variation as well as signal energy transformation.

This section includes four parts: (1) data comparison of different sampling points, (2) magnitude-based and energy-based DIs in the time domain, (3) amplitude-based and energy-based DIs in the frequency domain, and (4) signal fusion for all the actuator–sensor paths.

5.1. Data Comparison of Different Sampling Points

In general, it is well accepted that the more sampling points are acquired at an active sensor network, the better the signals are measured in an accurate, stable, and practicable manner. Different sampling points signals of S0 mode are extracted to calculate the magnitude/amplitude-based DIs and energy-based DIs in the time-frequency domain. We only select the representative path such as actuator–sensor path A_1–S_4, as shown in Figures 5 and 6.

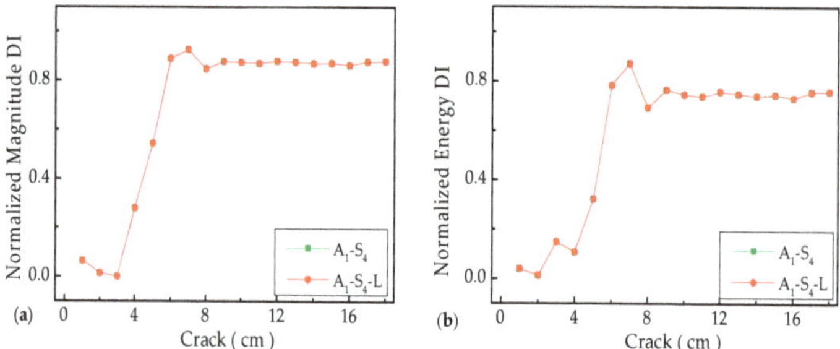

Figure 5. The magnitude and energy damage indexes (DIs) of different sampling points in the time domain (**a**) Normalized magnitude-based DIs; (**b**) Normalized energy-based DIs.

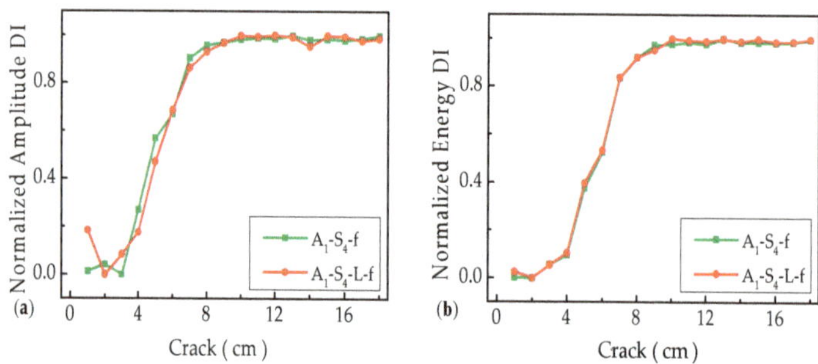

Figure 6. The amplitude and energy DIs of different sampling points in the frequency domain (**a**) Normalized amplitude-based DIs; (**b**) Normalized energy-based DIs.

A_1-S_4 stands for the sampling points of the common length in the time domain, and A_1–S_4–L refers to the sampling points of the longer length from the time domain. A_1–S_4–f and A_1–S_4–L–f correspond to the common length and longer length in the frequency domain, respectively. In the time domain, the magnitude-based and energy-based DIs of two different types of sampling points are exactly identical, as shown in Figure 5. Based on the amplitude-based DIs in the frequency domain, it is found that almost no fluctuations of DIs are observed for the reason that the measured signals are prone to be contaminated with lower noise and improved reliability in practice. Moreover, ultrasonic guided waves are highly dispersive and multimodal, which causes some difficulties in the process of signal feature extraction. Nevertheless, the overall distribution of magnitude/amplitude-based and energy-based DIs are well consistent. Considering that the differences between the magnitude/amplitude-based and energy-based DIs of different sampling points are very small, and they can be ignorable, to some extent. Therefore, the sampling points of the common length are selected to analyze and process the acquired signals in the following studies.

5.2. Magnitude-Based and Energy-Based DIs in Time Domain

The intensity of damage-reflected wave energy is taken as a sensitive indicator to determine the orientation of fatigue cracks, as correlated in which the strongest reflection is captured perpendicular to the direction of incidence. With the increase of the incidence angle of the guided waves, the intensity of the crack-reflected signal energy declines quickly. Based on this, the actuator–sensor paths are selected by extracting signal features associated with higher intensity energy [30], the gained signals also have a higher signal-to-noise ratio. Therefore, the investigations successively select the actuator–sensor paths A_1–S_4, A_2–S_5, A_3–S_6, A_2–S_4, and A_2–S_6 as representative paths in the time-frequency domain, as displayed in Figures 7–10. On this condition, the different paths A_1–S_4, A_2–S_5, and A_3–S_6 are perpendicular to the crack orientation with a fixed crack length of 1.0 cm. Similarly, A_2–S_4, A_2–S_5, and A_2–S_6 are selected as the comparison paths of A_1–S_4, A_2–S_5, and A_3–S_6, the magnitude variations of the extracted S0 mode in the time domain are presented in Figure 7.

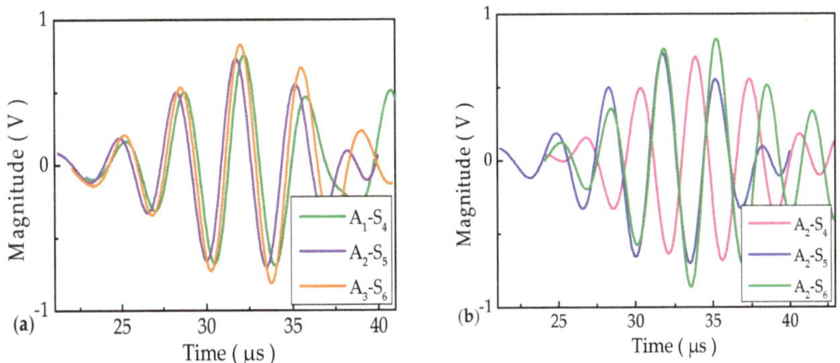

Figure 7. The magnitude of measured signals with different actuator–sensor paths in the time domain (a) A_1–S_4, A_2–S_5, and A_3–S_6 paths; (b) A_2–S_4, A_2–S_5, and A_2–S_6 paths.

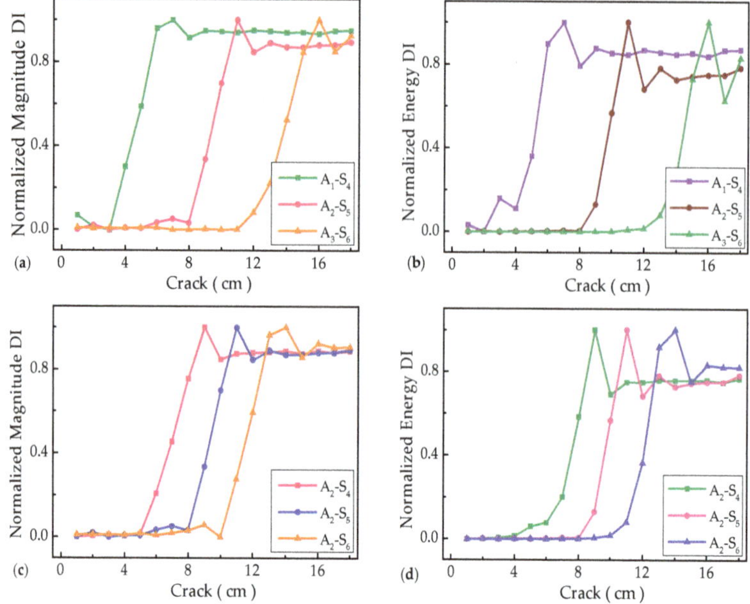

Figure 8. The magnitude and energy DIs of different actuator–sensor paths in the time domain (**a**) Normalized magnitude-based DIs of the A_1-S_4, A_2-S_5, and A_3-S_6 paths; (**b**) Normalized energy-based DIs of the A_1-S_4, A_2-S_5, and A_3-S_6 paths; (**c**) Normalized magnitude-based DIs of the A_2-S_4, A_2-S_5, and A_2-S_6 paths; (**d**) Normalized energy-based DIs of the A_2-S_4, A_2-S_5, and A_2-S_6 paths.

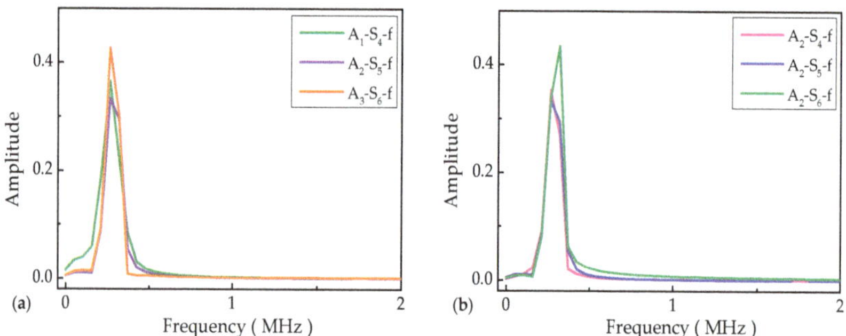

Figure 9. The amplitude of signals with different actuator–sensor paths in the frequency domain (**a**) A_1-S_4, A_2-S_5, and A_3-S_6 paths; (**b**) A_2-S_4, A_2-S_5, and A_2-S_6 paths.

The normalized magnitude-based and energy-based DIs in the time domain are calculated by Equations (2) and (4). As shown in Figure 8, in cases where the crack of the A_1-S_4 actuator–sensor path propagates from 0.0 cm to 3.0 cm, the A_2-S_5 actuator–sensor path propagates from 0.0 cm to 8.0 cm, or the A_3-S_6 actuator–sensor path propagates from 0.0 cm to 12.0 cm, there are no obvious variations in the DIs with some negligible fluctuations. The detection features of magnitude-based and energy-based DIs are substantially restricted to evaluate the fatigue crack, in case that the crack size is comparable with the probing wavelength. It is also found that the temporal features of guided waves are fairly limited to characterize the fatigue crack growth near the actuator–sensor paths with high sensitivity. When the progressive crack is large enough or near the actuator–sensor paths, with an increase of the fatigue crack, the remarkable DIs can be obtained. On the contrary, when the extended fatigue crack locates far from the actuator–sensor paths, it shows that the DIs are less sensitive to characterizing the crack variations without significant changes.

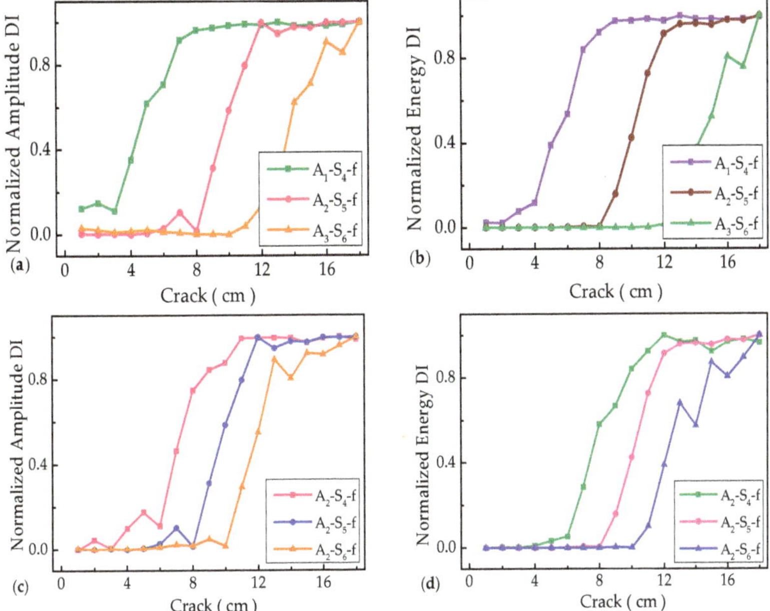

Figure 10. The amplitude and energy DIs of different actuator–sensor paths in the frequency domain (**a**) Normalized magnitude-based DIs of A_1–S_4, A_2–S_5, and A_3–S_6 paths; (**b**) Normalized energy-based DIs of A_1–S_4, A_2–S_5, and A_3–S_6 paths; (**c**) Normalized magnitude-based DIs of A_2–S_4, A_2–S_5, and A_2–S_6 paths; and (**d**) Normalized energy-based DIs of A_2–S_4, A_2–S_5, and A_2–S_6 paths.

5.3. Amplitude-Based and Energy-Based DIs in Frequency Domain

The normalized magnitude-based and energy-based DIs in the time domain can provide convenient conditions for crack propagation quantification, but their effectiveness in depicting crack size and damage level are still controversial. This is because small-scale cracks can cause a feeble variety for the crack-scattered wave in the time domain, that is, the magnitude of fatigue crack is often altered at an imperceptible level within the temporal information. Therefore, the signal features associated with the frequency domain are extracted to improve the sensitivity and stability of amplitude-based and energy-based DIs, in accordance with a pitch–catch sensor network configuration [25]. Fast Fourier transform (FFT) is used to process time-domain signals to obtain a frequency spectrum [31], as shown in Figure 9. Compared with the magnitude variations in the time domain from Figure 7, the amplitude variations in the frequency domain are illustrated in a more obvious and reliable manner.

The signals shown in Figure 8 are subsequently preformed with fast Fourier transform (FFT) to obtain the corresponding frequency spectrogram, as shown in Figure 10. The overall trend of DIs obtained in the frequency domain is basically consistent with that of the time domain. However, when the crack of the A_1–S_4 actuator-sensor path spreads from 0.0 cm to 6.0 cm, the A_2–S_5 actuator–sensor path spreads from 0.0 cm to 11.0 cm, or the A_3–S_6 actuator–sensor path spreads from 0.0 cm to 15.0 cm, there are no outstanding peaks for the amplitude-based and energy-based DIs. It is clearly observed that the variations of DIs are gradually and steadily obverted to the stable stage. In addition, the variations of DIs can be regarded as insignificant in the stable stage, which are consistent with the observations in Figure 8.

5.4. Signal Fusion for All Actuator–Sensor Paths

In order to strengthen the crack-interrelated features of the individual original DIs, the environmental noise and measurement uncertainties (unwanted information in individual original DIs) need to be lessened effectively, and the signal-to-noise ratio of the captured crack signal features need to be heightened as much as possible. According to the DI weights of each actuator–sensor path, the extracted original DIs are required to decentralize individual contributions equally and amalgamate all available signal features at each specific crack. A synthetic DI related with fatigue crack is proposed, which can be used to characterize the crack propagation at all paths. Based on the analysis in the time-frequency domain, the hybrid magnitude/amplitude-based DIs (defined by Equations (6) and (7)) and energy-based DIs (defined by Equations (8) and (9)) calculated by the fusion arithmetic are separately illustrated in Figures 11 and 12.

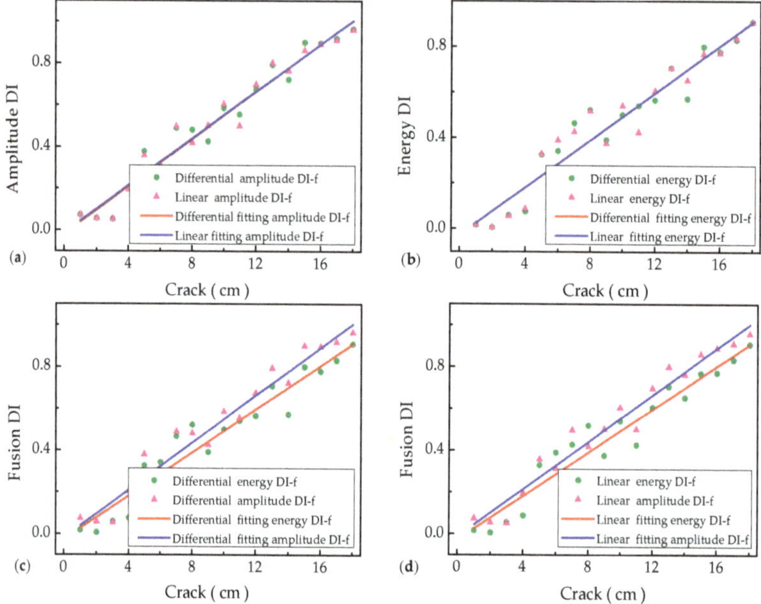

Figure 11. The magnitude and energy DIs of different fusion methods in the time domain (**a**) Linear and differential fusion magnitude DIs; (**b**) Linear and differential fusion energy DIs; (**c**) Differential fusion of magnitude-energy DIs; and (**d**) Linear fusion of magnitude-energy DIs.

Firstly, the linear fitting of DIs is conducted at different crack growth regions for each actuator–sensor path. Based on the DI weights of different paths, linear fusion DIs are extracted for all the actuator–sensor paths. The first-order differential is the difference between two consecutive neighbors in a discrete DI function. Differential fusion DIs are obtained by different weights of all the paths. Based on the analysis of linear fusion DIs and differential fusion DIs (including magnitude/amplitude-based and energy-based DIs), Figure 11 severally displays the ultimate DIs in the time domain. Experimental results show that the propagation of the fatigue crack can be calibrated by the synthetic DIs quantitatively. It is found that the fitting differences between the linear magnitude and differential magnitude DIs are especially tiny, and the fitting straight lines are mostly coincident. Moreover, it can also be clearly observed that the stability of linear and differential magnitude fitting DIs is better than that of linear energy fitting and differential energy fitting DIs. Simultaneously, the results indicate that the stabilities of differential energy fitting DIs are the worst in the time domain.

In other words, the fitting effectiveness from different fitting methods can be represented alternatively by the fitting correlation coefficients and mean squared error in Tables 3 and 4.

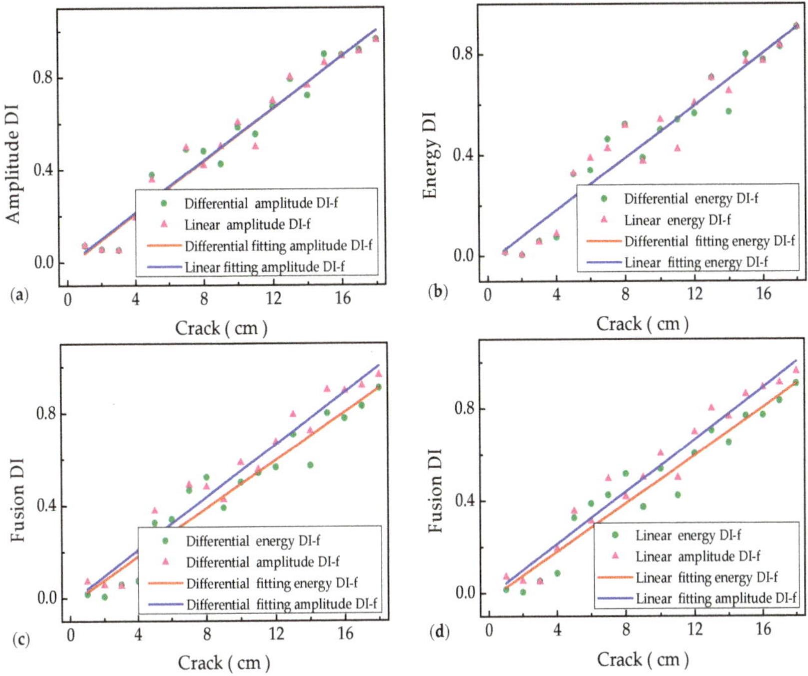

Figure 12. The amplitude and energy DIs of different fusion methods in the frequency domain (**a**) Linear and differential fusion amplitude DIs; (**b**) Linear and differential fusion energy DIs; (**c**) Differential fusion of amplitude-energy DIs; (**d**) Linear fusion of amplitude-energy DIs.

Table 3. The correlation coefficients of various acoustic features.

Correlation Coefficients	Time Domain Analysis	Frequency Domain Analysis
Linear magnitude/amplitude	0.97529	0.9806
Differential magnitude/amplitude	0.97284	0.97793
Linear energy	0.94154	0.96643
Differential energy	0.85893	0.96573

Table 4. The mean squared error of various acoustic features.

Mean Squared Error	Time Domain Analysis	Frequency Domain Analysis
Linear magnitude/amplitude	0.00682	0.00384
Differential magnitude/amplitude	0.00554	0.00445
Linear energy	0.01491	0.00577
Differential energy	0.0064	0.00589

Similarly, the fitting DIs from the frequency domain can characterize the propagation of crack accurately, as shown in Figure 12. Obviously, it indicates that the overall fitting DIs in the frequency domain coincide with the fitting in the time domain. Whereas, it is corroborated that the stability and reliability of the fitting DIs in the frequency domain are better than those in the time domain. Simultaneously, it is particularly interesting that small differences between the experimental measurement DIs and the fitting DIs can be clearly identified in Figure 12, Tables 3 and 4.

It can be concluded that the synthetic fusion DIs of different acoustic features can be used to calibrate the propagation of fatigue crack quantitatively to a certain extent. However, considering the stability, reliability, and practicability of the signals, the linear and differential amplitude fusion schemes in the frequency domain are shown to have better promising alternatives for characterizing the crack propagation.

6. Conclusions

It is of great significance but also a highly challenging assignment to monitor small-scale fatigue crack continuously. In addition, it is of vital importance for calibrating the fatigue crack propagation quantitatively, synchronously. In this paper, the typical ultrasonic guided waves signal characteristics in the time-frequency domain are acquired from an SMA490BW steel plate associating with an active sensor network. The crack characterization approaches using various acoustic features (capitalizing on the magnitude/amplitude-based and energy-based DIs in the time-frequency domain) are established by constructing different DIs from the extracted S0 mode signals of ultrasonic guided waves, which are used for online monitoring of the fatigue crack propagation quantitatively. The results reveal that the magnitude/amplitude-based and energy-based DIs of individual actuator–sensor paths are substantially restricted to characterize the fatigue crack with a size on the same order of the magnitude of the central wavelength or indicate fatigue crack propagation near the paths with high sensitivity. Subsequently, according to the weights of individual paths, a synthetic fusion scheme is developed to fuse all available DIs for all paths. Using the linear fused DIs and differential fused DIs in the time-frequency domain, a synthetic fusion DI scheme of different acoustic features is proposed, which can indicate the propagation of fatigue crack accurately. It is also clear that the stability and reliability of the fused DIs calculated by the acoustic features in the frequency domain are better than those in the time domain. Ultimately, it is important to realize that the linear and differential amplitude fusion DIs of a bogie plate-like structure in the frequency domain are more promising to characterize the crack propagation quantitatively than other fused ones.

Author Contributions: Conceptualization, H.J.; Data curation, H.J. and J.Y.; Formal analysis, H.J.; Methodology, H.J., W.L. and X.Q.; Project administration, X.Q.; Supervision, W.L.; Validation, H.J.; Writing—original draft, H.J.; Writing—review & editing, H.J., W.L. and X.Q.

Funding: This research was funded by the National Natural Science Research Foundation of China (Grant Numbers 11972314, 11774295) and Jiangsu Bide Science and Technology Co. Ltd.

Conflicts of Interest: The authors declare no conflict of interest.

References

1. Campbell, F.C. *Elements of Metallurgy and Engineering Alloys*; ASM International: Cleveland, OH, USA, 2008.
2. Kulkarni, S.S.; Achenbach, J.D. Structural health monitoring and damage prognosis in fatigue. *Struct. Health Monit.* **2008**, *7*, 37–49. [CrossRef]
3. Su, Z.; Zhou, C.; Hong, M.; Cheng, L.; Wang, Q.; Qing, X. Acousto-ultrasonics-based fatigue damage characterization: Linear versus nonlinear signal features. *Mech. Syst. Signal Proc.* **2014**, *45*, 225–239. [CrossRef]
4. Yang, B.; Zhao, Y.X. Experimental research on dominant effective short fatigue crack behavior for railway LZ50 axle steel. *Int. J. Fatigue* **2012**, *35*, 71–78. [CrossRef]
5. Zhao, Y.X.; Yang, B. Scale-induced effects on fatigue properties of the cast steel for Chinese railway rolling wagon bogie frames. *Adv. Mater. Res.* **2010**, *118–120*, 59–64. [CrossRef]
6. Kundu, T.; Das, S.; Jata, K.V. Health monitoring of a thermal protection system using Lamb waves. *Struct. Health Monit.* **2008**, *8*, 29–45. [CrossRef]
7. Sohn, H.; Lee, S.J. Lamb wave tuning curve calibration for surface-bonded piezoelectric transducers. *Smart Mater. Struct.* **2009**, *19*, 15007–15012. [CrossRef]
8. Mitra, M.; Gopalakrishnan, S. Guided wave based structural health monitoring: A review. *Smart Mater. Struct.* **2016**, *25*, 053001. [CrossRef]

9. Hong, M.; Su, Z.; Lu, Y.; Sohn, H.; Qing, X. Locating fatigue damage using temporal signal features of nonlinear Lamb waves. *Mech. Syst. Signal Proc.* **2015**, *60*, 182–197. [CrossRef]

10. Lim, H.J.; Sohn, H.; Kim, Y. Data-driven fatigue crack quantification and prognosis using nonlinear ultrasonic modulation. *Mech. Syst. Signal Proc.* **2018**, *109*, 185–195. [CrossRef]

11. Qing, X.; Li, W.; Wang, Y.; Sun, H. Piezoelectric transducer-based structural health monitoring for aircraft applications. *Sensors* **2019**, *19*, 545. [CrossRef]

12. Qing, X.P.; Beard, S.J.; Kumar, A.; Ooi, T.K.; Chang, F.K. Built-in sensor network for structural health monitoring of composite structure. *J. Intel. Mater. Syst. Struct.* **2006**, *18*, 39–49. [CrossRef]

13. Xie, F.; Li, W.; Zhang, Y. Monitoring of environmental loading effect on the steel with different plastic deformation by diffuse ultrasound. *Struct. Health Monit.* **2018**, *18*, 602–609. [CrossRef]

14. Ostachowicz, W.; Kudela, P.; Malinowski, P.; Wandowski, T. Damage localisation in plate-like structures based on PZT sensors. *Mech. Syst. Signal Proc.* **2009**, *23*, 1805–1829. [CrossRef]

15. Kundu, T.; Nakatani, H.; Takeda, N. Acoustic source localization in anisotropic plates. *Ultrasonics* **2012**, *52*, 740–746. [CrossRef]

16. Michaels, J.E. Detection, localization and characterization of damage in plates with an in situ array of spatially distributed ultrasonic sensors. *Smart Mater. Struct.* **2008**, *17*, 035035. [CrossRef]

17. Koh, Y.L.; Chiu, W.K.; Rajic, N. Effects of local stiffness changes and delamination on Lamb wave transmission using surface-mounted piezoelectric transducers. *Compos. Struct.* **2002**, *57*, 437–443. [CrossRef]

18. Wu, Z.; Qing, X.P.; Chang, F.K. Damage detection for composite laminate plates with a distributed hybrid PZT/FBG sensor network. *J. Intel. Mat. Syst. Struct.* **2009**, *20*, 1069–1077.

19. Ramadas, C.; Balasubramaniam, K.; Joshi, M.; Krishnamurthy, C.V. Interaction of guided Lamb waves with an asymmetrically located delamination in a laminated composite plate. *Smart Mater. Struct.* **2010**, *19*, 065009. [CrossRef]

20. Li, D.; Kuang, K.S.C.; Koh, C.G. Fatigue crack sizing in rail steel using crack closure-induced acoustic emission waves. *Meas. Sci. Technol.* **2017**, *28*, 065601. [CrossRef]

21. Gagar, D.; Foote, P.; Irving, P. A novel closure based approach for fatigue crack length estimation using the acoustic emission technique in structural health monitoring application. *Smart Mater. Struct.* **2014**, *23*, 105033. [CrossRef]

22. Ihn, J.B.; Chang, F.K. Detection and monitoring of hidden fatigue crack growth using a built-in piezoelectric sensor/actuator network: II. Validation using riveted joints and repair patches. *Smart Mater. Struct.* **2004**, *13*, 621–630. [CrossRef]

23. Masserey, B.; Fromme, P. Fatigue crack growth monitoring using high-frequency guided waves. *Struct. Health Monit.* **2013**, *12*, 484–493. [CrossRef]

24. Qiu, L.; Yuan, S.; Bao, Q.; Mei, H.; Ren, Y. Crack propagation monitoring in a full-scale aircraft fatigue test based on guided wave-Gaussian mixture model. *Smart Mater. Struct.* **2016**, *25*, 055048. [CrossRef]

25. Zhou, C.; Su, Z.; Cheng, L. Probability-based diagnostic imaging using hybrid features extracted from ultrasonic Lamb wave signals. *Smart Mater. Struct.* **2011**, *20*, 125005. [CrossRef]

26. Migot, A.; Bhuiyan, Y.; Giurgiutiu, V. Numerical and experimental investigation of damage severity estimation using Lamb wave-based imaging methods. *J. Intel. Mater. Syst. Struct.* **2019**, *30*, 618–635. [CrossRef]

27. Li, W.; Cho, Y.; Achenbach, J.D. Detection of thermal fatigue in composites by second harmonic Lamb waves. *Smart Mater. Struct.* **2012**, *21*, 85–93. [CrossRef]

28. Yan, J.; Jin, H.; Sun, H.; Qing, X. Active Monitoring of Fatigue Crack in the Weld Zone of Bogie Frames Using Ultrasonic Guided Waves. *Sensors* **2019**, *19*, 3372. [CrossRef] [PubMed]

29. Zhu, J.; Wang, Y.; Qing, X. Modified electromechanical impedance-based disbond monitoring for honeycomb sandwich composite structure. *Compos. Struct.* **2019**, *217*, 175–185. [CrossRef]

30. Zhou, C.; Su, Z.; Cheng, L. Quantitative evaluation of orientation-specific damage using elastic waves and probability-based diagnostic imaging. *Mech. Syst. Signal Proc.* **2011**, *25*, 2135–2156. [CrossRef]

31. Magdalena, P. Spectral Methods for Modelling of Wave Propagation in Structures in Terms of Damage Detection—A Review. *Appl. Sci.* **2018**, *8*, 1124.

© 2019 by the authors. Licensee MDPI, Basel, Switzerland. This article is an open access article distributed under the terms and conditions of the Creative Commons Attribution (CC BY) license (http://creativecommons.org/licenses/by/4.0/).

Article

Microscopic Multiple Fatigue Crack Simulation and Macroscopic Damage Evolution of Concrete Beam

Baijian Wu [1,2], **Zhaoxia Li** [1,2], **Keke Tang** [3,*] **and Kang Wang** [2]

[1] Jiansu Key Laboratoryof Engineering Mechanics, Southeast Uiversity, Nanjing 210096, China; bawu@seu.edu.cn (B.W.); zhxli@seu.edu.cn (Z.L.)

[2] Department of Engineering Mechanics, Southeast University, Nanjing 210096, China; 220120870@seu.edu.cn

[3] School of Aerospace Engineering and Applied Mechanics, Tongji University, Shanghai 200092, China

* Correspondence: kktang@tongji.edu.cn

Received: 5 October 2019; Accepted: 31 October 2019; Published: 1 November 2019

Abstract: Microcracks in concrete can coalesce into larger cracks that further propagate under repetitive load cycles. Complex process of crack formation and growth are essentially involved in the failure mechanism of concrete. Understanding the crack formation and propagation is one of the core issues in fatigue damage evaluation of concrete materials and components. In this regard, a numerical model was formulated to simulate the thorough failure process, ranging from microcracks growth, crack coalescence, macrocrack formation and propagation, to the final rupture. This model is applied to simulate the fatigue rupture of three-point bending concrete beams at different stress levels. Numerical results are qualitatively consistent with the experimental observations published in literature. Furthermore, in the framework of damage mechanics, one damage variable is defined to reflect stiffness reduction caused by fatigue loading. S-N curve is subsequently computed and the macroscopic damage evolution of concrete beams are achieved. By employing the combined approaches of fracture mechanics and damage mechanics, made possible is the damage evolution of concrete beam as well as the microscopic multiple fatigue crack simulation. The proposed approach has the potential to be applied to the fatigue life assessment of materials and components at various scales in engineering practice.

Keywords: microcracks; multiple fatigue crack; crack coalescence; concrete beams; damage evolution

---

1. Introduction

Fatigue problems are prevalent in the service life of concrete structures such as bridge slabs, highway pavements and offshore structures [1–3]. Concrete lies in the category of quasi-brittle material and it is microscopically heterogeneous. Combination of concrete and other materials are extensively applied in engineering practice [4–6]. A large number of flaws are inevitably created before the loading, which can be considerably attributed to the loss of concrete moisture [7]. Under repetitive fatigue load cycles, pre-existing flaws or microcracks in concrete can trigger the initiation, coalesce into larger cracks that further propagate, ultimately leading to the final rupture. The so-called fatigue damage accumulation is a progressive, permanent and localized internal changes in the concrete. Complex process of crack formation and growth are essentially involved in the failure mechanism of concrete. Understanding the crack formation and propagation is one of the core issues in fatigue damage evaluation of concrete materials and components. How crack progression throughout the fatigue life affects and causes the final failure modes remains to be clarified.

The approaches of damage mechanics and fracture mechanics are generally employed to model the progressive fatigue accumulation in concrete materials [8–12]. In the framework of damage mechanics, a damage variable is required to be defined. A physically reasonable damage evolution law needs to be formulated such that the progressive material degradation caused by microcrack initiation, coalescence and propagation is reflected through the process. Both the microcrack evolution and stiffness reduction simultaneously reflect the material deterioration, microscopically and macroscopically. It should be pointed out that, in the concepts of damage mechanics, it is assumed that there are no pre-existing microcracks or flaws in material specimens. Damage evolution is a continuous process that describes damage or strain localization and characterizes the fatigue behavior of concrete, though microcracks are randomly distributed in specimens.

A series of fatigue models have been formulated based on thermodynamics concepts in the framework of damage mechanics [13,14]. Progressive stiffness reduction is straightforwardly reflected in these models. Nevertheless, most of these models are empirical and lack of physical basis. Damage parameters are highly dependent on experimental data. In recent years, state-of-the-art experimental technology has been introduced to this field. By using industrial computed tomography (ICT) technology, spatial distribution of fatigue cracks within concrete is observed. Subsequently a macroscopic fatigue damage parameter D is subsequently defined based on the quantitative description method of the overall distribution of fatigue cracks in concrete [15]. This combined micro–macro approach can better characterize the evolution process of material fatigue damage of concrete under compression from the overall distribution of fatigue cracks. A refined engineering rule for the assessment of remaining fatigue life of concrete under compressive cyclic loading with varying amplitude is proposed in [16]. The proposed empirical rule is derived based on a combined numerical and experimental investigation of the loading sequence effect. The equivalent tensile strain rate is adopted to govern the fatigue damage evolution in the applied modeling approach. It is generally accepted that models derived from physical principles are relatively reliable compared to the empirical models. These physical laws include energy principles, dimensional analysis as well as similitude concepts. It is noteworthy, a fatigue model for plain concrete under variable amplitude loading is proposed in [17] by unifying the concepts of damage mechanics and fracture mechanics through an energy equivalence. The whole work is on the basis of thermodynamic framework using the principles of dimensional analysis and self-similarity. A closed form expression for dual dissipation potential is derived. A damage evolution law is further proposed to compute damage in the volume element subjected to fatigue loading. This proposed model incorporates the complex behavior of concrete under fatigue and provides a more rational method for fatigue life evaluation of concrete materials and components. Some other physically based constitutive modeling of concrete fatigue can be found in [9,14,18].

Fracture mechanics serves the ideal mechanistic tool for concrete fatigue damage evolution, considering the inherent microcracks and flaws in concrete. Fracture mechanics approach is also extensively applied to crack initiation, crack growth rate, crack density evolution and stiffness degradation in other materials such as composite laminates [19–21]. The fatigue models in fracture mechanics are mostly the empirical Paris type equation, in which the crack growth rate with respect to number of fatigue load cycles is correlated to the stress intensity factor range. Not surprisingly, the Paris law is empirically formulated on the basis of metallic fatigue. Efforts have been taken to modify the Paris law by introducing extra crack growth influencing parameters [22,23]. Based on previous models, a newly proposed fatigue crack propagation model for concrete beam incorporates the effect of loading frequency of applied load, loading history as well as size effect parameters. An analytical crack growth model is developed to predict fatigue behavior of quasi-brittle materials in [24]. Early work related to concrete fatigue implies that parameters in the classical power law are dependent on micro-structural size, crack size and size scale [23]. An improved crack propagation model for plain concrete under fatigue loading is derived in [25], considering the effect of critical energy dissipation in fatigue. The so-called fatigue fracture energy is able to capture the observed size effect in concrete fatigue. The model is analytical and influence of fracture process zone is incorporated in the proposed formulation. An irreversible cohesive zone model for interface fatigue crack growth simulation is developed in [26]. The improved CZM is physically based and the traction–separation behavior does not follow a predefined approach. This proposed model for the computational simulation of

FCG provides a crucial step in the direction of mechanistic mode developments in the area of fatigue crack growth. A predictive cohesive modeling framework for corrosion fatigue is proposed in [27]. These CZMs could also offer reference for concrete fatigue modeling.

In recent years, it is noteworthy that efforts are particularly being devoted to both microscopic and macroscopic aspects [28–30]. A microplane constitutive damage model is developed in [28]. The proposed model is able to describe both the fatigue crack growth and the nonlinear triaxial damage behavior of concrete. A discrete element modeling approach for fatigue damage growth in cemented materials are developed in [29]. The model formulation is based on coupling damage mechanics and plasticity theory and combining with a fatigue damage evolution law to describe the degrading response of cemented materials subjected to cyclic loading. Global fatigue damage response as well as microstructural effects could be reflected. A physical stochastic damage model for concrete subjected fatigue loading is formulated in [30]. It could be put in the framework of mesoscopic stochastic fracture models that are capable of reflecting the general nonlinearity and randomness in the mechanical behavior of concrete. It is fair to say, our work is motivated by the aforementioned efforts.

In this context, the purpose of approaches related to damage or fracture mechanics is to develop effective models that are able to reflect the progressive material degradation under fatigue loading. It shall be pointed out that the model is stipulated to pure concrete materials, the internal longitudinal and transverse reinforcements with related effects [4–6] are not considered in the present work. Based on these proposed models, attempts have been particularly made to the numerical simulation of fatigue [18,31]. The advantage of implementing a developed fatigue model in available finite element codes is that both constitutive equations and interface models can be included in the numerical simulation. The fatigue and mechanical behaviors of concrete materials and components are predicted through the numerical implementation in commercial FE software. A numerical procedure is developed to simulate crack propagation behaviors at the concrete aggregate-matrix interface in [31]. It shall be pointed out that, modeling of single aggregate-matrix interface crack is studied. Based on the conclusions in [31], it is assumed that interface crack can be mapped into idealized crack along the main axis of aggregate to solve the multiple crack coalescence and propagation in concrete. This leads to the work described in [32]. On the basis of these work, microscopic multiple fatigue crack simulation and macroscopic damage evolution of concrete beam subjected to cyclic loads is investigated in the present work. We are mainly focused on the fatigue damage analysis microscopically and macroscopically. In this regard, the aim of the present work is to define a damage variable that reflect the fatigue evolution of concrete specimens, on the basis of the multiple crack simulation in concrete beams.

The work presented in this paper is structured as follows: described in Section 2 are the numerical model for multiple crack simulation and fatigue crack growth law. Numerical modeling of three-point bending beams are presented in Section 3. Macroscopic fatigue damage analysis is discussed in Section 4. The main conclusions are summarized in Section 5.

2. Microscopic Multiple Crack Simulation and Fatigue Crack Growth Law

At early age, initial microcracks usually emerge at the interface of cement paste and aggregate due to the natural shrinkage combined with other factors at the time of curing. The crack could propagate along the interface or penetrate into matrix or aggregate. In our previous work [31], modeling of one crack propagation in concrete is made possible (Figure 1). The multiple-phase structure of concrete material could be simplified as a continuum with multiple microcracks. Here we are mainly focused on simulation of multiple cracks as well as their parallel growths under fatigue loads. Figure 2 shows an example of the multiple crack model for concrete material.

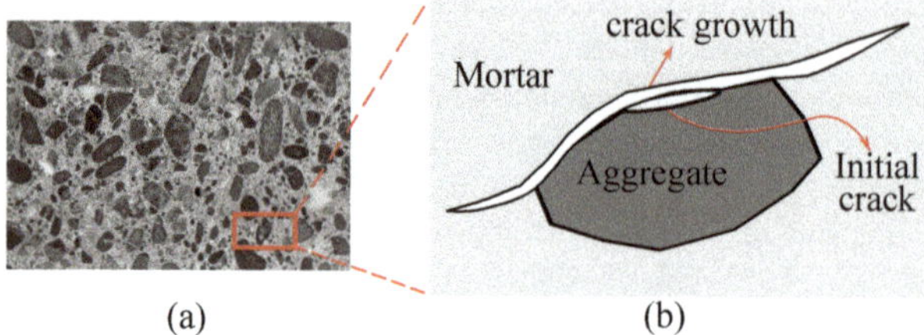

Figure 1. Schematic of single crack growth in concrete. (**a**) Schematic of concrete material; (**b**) growth of single microcrack.

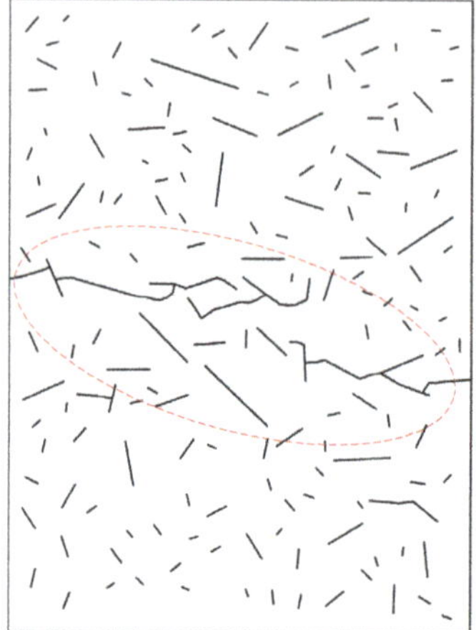

Figure 2. Multiple microcrack growth and linking: an example.

2.1. Modeling of Multiple Crack Growth

When a number of microcracks start to propagate simultaneously, induced by external fatigue load, the morphology of crack linking with each other may be very complex. However, it could be decomposed into three basic modes. As shown in Figure 3a, the two endpoints of Crack 1 and 2 meet together such that the two cracks merge into a single crack. Figure 3b shows Crack 4 is growing intersecting with Crack 3. In this situation, stress singularity at the intersected endpoint of Crack 4 vanishes—this crack tip will stop growing. Therefore, the situation of Figure 3c will not happen in the present work, though it could be simulated as well.

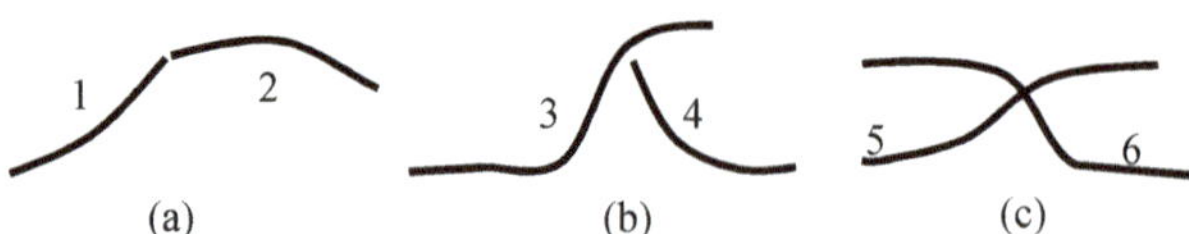

Figure 3. Basic modes of crack linking. (**a**) meeting; (**b**) merging; (**c**) intersecting.

In the present study, a program has been developed comprehensively to achieve all crack growing situations, including detecting and treating crack merging and intersecting, multiple crack growth, and detecting boundary edges [32]. Presented in the highlighted area of Figure 2 is a relatively complicated situation of crack growth and multiple crack linking. It could be seen that the method could successfully simulate the concurrent growth of microcracks, crack coalescence, the formulation and growth of macrocracks, and the final rupture. Details of the approach can be found in [31,32].

On the basis of the formulated multiple crack simulation, we are mainly focused on the fatigue crack growth in concrete materials.

2.2. Fatigue Crack Growth Law

The fatigue crack propagation law is formulated by Paris and Erdogan based on experiments [33], which points out the relationship between stress intensity factor and crack growth rate, namely

$$\frac{da}{dN} = C(\Delta K)^m. \tag{1}$$

Here da/dN refers to the crack growth rate. a is the current crack length, and ΔK is the incremental stress intensity factor. C and m are the empirical fatigue constants obtained from material tests. In this paper, a modified Paris law is introduced to govern the growth behavior of microcracks. The whole crack growth process is divided into three stages as displayed in Figure 4.

1. In the low rate zone (Stage I) $\Delta K < \Delta K_{th,1}$, the crack is set not to propagate because the rate is very low. The fatigue threshold is chosen as

$$\Delta K_{th1,J} = 0.2 K_{C,J}. \tag{2}$$

2. In the medium rate zone (Stage II) $\Delta K > \Delta K_{th,1}$ and $\Delta K < \Delta K_{th,2}$, the crack growth follows the Paris law [33].

$$\frac{da_J}{dN} = C(\Delta K_J)^m. \tag{3}$$

   According to the experiments, the value of m for concrete is generally between 1.0 and 2.5. In order to make the crack growth rate lies between 10^{-5} mm/c and 10^{-2} mm/c [12,34], the empirical parameters C and m are adopted as $C = 10^{-8}$, $m = 1.5$.

3. In the high rate zone (Stage III) $\Delta K > \Delta K_{th,2}$, the crack will quickly propagate in an unstable fashion, finally lead to rupture. For the convenience of numerical simulation, here in this zone, the growth rate is practically set to be 1.2×10^{-3} mm/c. Also, the 2nd fatigue threshold $\Delta K_{th,2}$ is set to be

$$\Delta K_{th2,J} = 0.8 K_{C,J}. \tag{4}$$

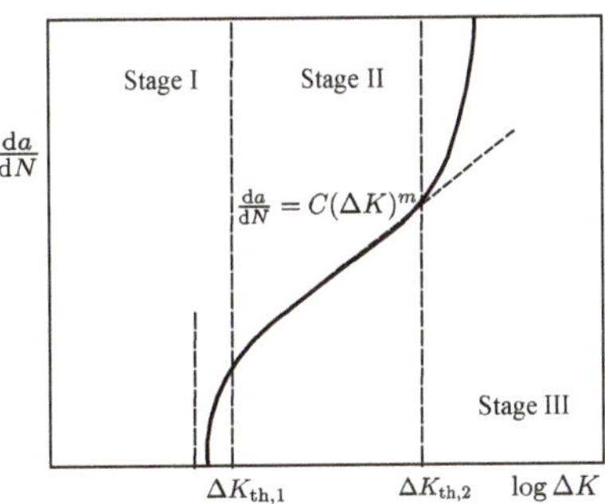

Figure 4. Fatigue crack growth rate.

Microcracks subjected to repetitive load, start to propagate, though advancement of some cracks can be ignored. Subsequently, these cracks may coalesce into several macrocracks and finally reach to the stage of final rupture. In the actual computing process, one 'computational step' is not equal to one 'load cycle step'. Since the crack shows a 'significant' growth only after the load has been repeated for a certain number of cycles, Δa-controlled rather than N-controlled strategy is used here. In each step, marked by superscript i, the growth length of the crack that has the highest rate, assuming to be the β^i-th crack, is controlled to be Δa^i. The number of load cycles ΔN^i is determined by the following formula

$$\Delta N^i = \frac{\Delta a^i}{C(\Delta K_\beta)^m}.$$ (5)

Then, incremental lengths for all cracks in this step could be calculated as

$$\Delta a_j^{(i)} = C(\Delta K_j^{(i)})^m \Delta N^{(i)}.$$ (6)

Summing up to get the total fatigue life

$$N = \sum \Delta N^i$$ (7)

Displayed in Figure 5 is the flowchart of numerical algorithm for multiple fatigue crack simulation. It shall be emphasized that both macroscopic and microscopic models are created such that a multi-scale model is further created via interface linking. Fatigue crack growth law is subsequently adopted to study the multiple microcrack growth, coalescence and macrocrack formation. After the final rupture is reached, result extraction and data analysis is made possible.

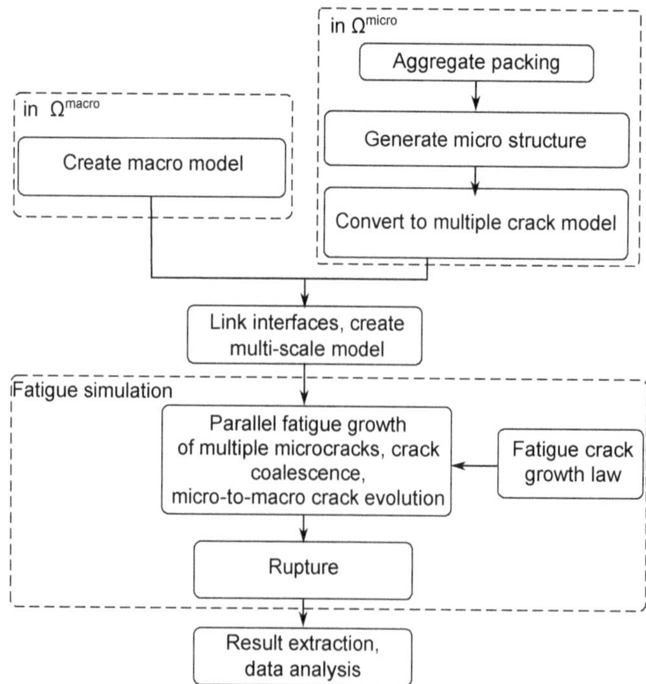

Figure 5. Algorithm for concrete fatigue simulation.

3. Fatigue Modeling of Three-Point Bending Beams

3.1. Three-Point Bending Beam Model

Beam is an important object in the study of concrete fatigue. Liner elastic constitution law is adopted for concrete. The model discussed in this paper is a three-point bending beam with a length of L, height of H and thickness of T. A concentrated force F is applied at the mid-span point, refer to Figure 6.

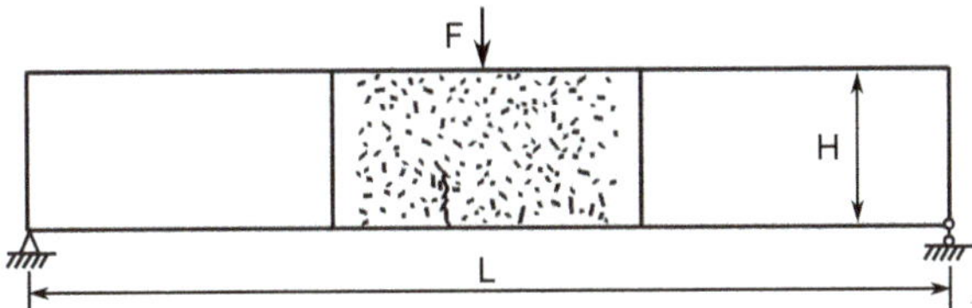

Figure 6. Model of a three-point bending concrete beam.

In order to improve the computational efficiency, the whole model is divided into two areas, the macro-scale area Ω^{Macro} (side area) and the micro-scale area Ω^{Micro} (middle area). The difference between the two regions is: (1) The macro-scale region is considered as homogeneous material, while the micro-scale region is heterogeneous, meaning that this area contains randomly distributed microcracks, (2) the typical element size d of the two regions is different, d^{Macro} is usually about 10 to 40 times of d^{Micro}. The element size of middle area is usually about 0.2–0.4 mm. Eight-node quadrilateral elements are subsequently generated.

In order to accurately simulate the stress field at the crack tip and to obtain stress intensity factor in the microscale area, singular elements needs to be added at crack tips, as shown in the Figure 7d. On the interface between these two regions, the connections are made by forcing the displacements to be equal. On the interface, the nodes of the coarse mesh are the master nodes, and the nodes of the fine mesh are the slave nodes. That is to say, for any microscopic node $A \in \partial\Omega^{\text{Micro}}$ locating at $\mathbf{x}$, its displacement u^{Micro} is forced to be

$$u^{\text{Micro}}(\mathbf{x}) = \sum_i N_i^{\text{Macro}}(\mathbf{x})u_i^{\text{Macro}}. \tag{8}$$

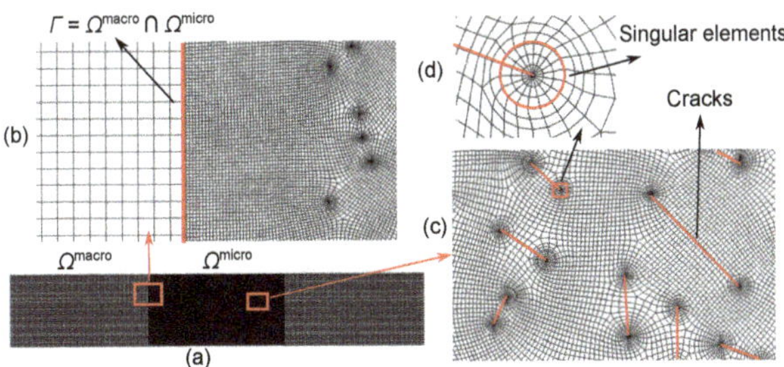

Figure 7. Multi-scale model of concrete beams. (**a**) Concrete model; (**b**) interface of macro-scale area and micro-scale area; (**c**) microcracks at micro-scale area; (**d**) singular elements.

Figure 8 shows a stress contour of a bending concrete beam at the stage of fatigue rupture. It could be seen that in a global sense, the numerical results agree with the corresponding conclusion from classical beam theory. Moreover, the stress level may fluctuate in the local area, which can be attributed to the microscopic heterogeneity in concrete material.

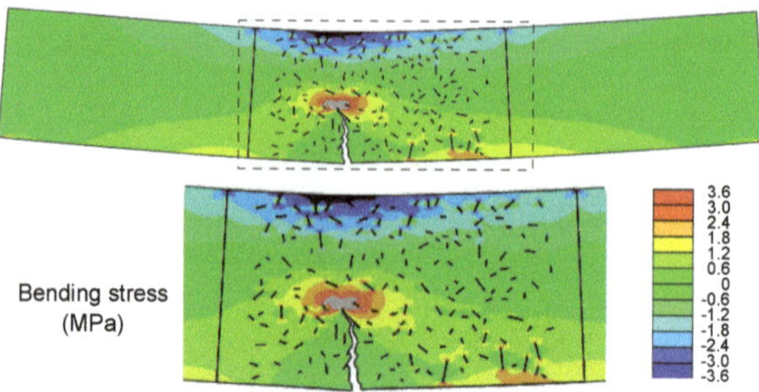

Figure 8. Stress contour of micro-scale area of concrete beam.

3.2. Case Design

As shown in Figure 6, beam height $H = 200$ mm, length $L = 800$ mm, The concentrated sinusoidal loading is located at the middle span, and the loading frequency is 10 Hz. The parameter

$$R = F_{\min}/F_{\max}$$

is chosen to be constant 0.1, where $F_{\min}$ and $F_{\max}$ denote the maximum and minimum force per cycle, respectively. Considering that

$$R = \frac{F_{\min}}{F_{\max}} = \frac{K_{J,\min}}{K_{J,\max}}, \tag{9}$$

We have

$$\Delta K_J = K_{J,\max}(1 - R). \tag{10}$$

That is to say, in each step, we can simply apply $F_{\max}$ on the beam to get ΔK_J at all crack tips. Also, stress level S is introduced as

$$S = \frac{F_{\max}}{F_u}. \tag{11}$$

Here F_u is the bending capacity of this beam. For three-point bending case, F_u could be estimated as 100 N via the formula

$$F_u = \sigma_u \frac{2T \cdot H^2}{3L}, \tag{12}$$

where σ_u is the tensile strength of concrete. In this paper, the value of S is tuned from 0.1 to 0.9, to investigate the influence of fatigue stress ratio.

3.3. Fatigue Failure Analysis

In order to perform fatigue failure analysis, it shall be defined that the ultimate fatigue load is reached once the nominal stiffness of the beam is only 5% left. The so-called macroscopic mechanical behaviors are substantially influenced by microscopic fatigue crack growth. A cycle-jump technique is employed to model load cycles such that the computational cost can be saved. In the simulations of regular cyclic loading scenarios with the same loading range, a cycle can be adopted to reduce the computational cost. As an example, the results at the stress level $S = 0.4$ are shown here. The three-point bending concrete beam is ruptured after 42,000 load cycles. Figure 9 shows crack propagation diagrams of the whole failure process, where N_f is the fatigue life of this specimen.

There are naturally a large number of distributed microscopic cracks in concrete, usually occurring on the mortar-aggregate interfaces during cement hydration process. Under the fatigue load, microcracks in tensile zone of the beam start to stably grow at a medium rate (Stage II) according to Paris law, as shown in Figure 9a,b. As the number of fatigue load increases, crack coalescence initially takes place at the lower left corner of the specimen. Localized damage subsequently appears.

It shall be pointed out that, crack coalescence does not necessarily occur in the designated area. The localized damage at the lower left corner of the specimen here is a reflection of random distribution of microcracks in concrete. To be able to describe this stochastic behavior is the strongest point of our algorithm. Once macrocrack appears, fatigue crack growth start to accelerate (but still in a stable fashion), gradually entering into Stage III, as shown in Figure 9c,d. The formation of macrocrack can be firstly seen in Figure 9c. Its further propagation will lead to final beam rupture. It should be noted that at this stage, cracks in other areas still grow at a low or medium rate. At later period of fatigue rupture, the neutral axis gradually moves up, the macrocrack on the left side propagates at a high rate, and the localized damage (on the upper left side) increases dramatically, which eventually leads to the rupture of beam, as shown in Figure 9e.

It is noteworthy that the unsymmetrical pattern of crack formation and propagation is clearly displayed in the middle area of concrete beam. It is reflective of the microscopically stochastic phenomenon of crack distribution. This is the strong point that our proposed model is able to achieve.

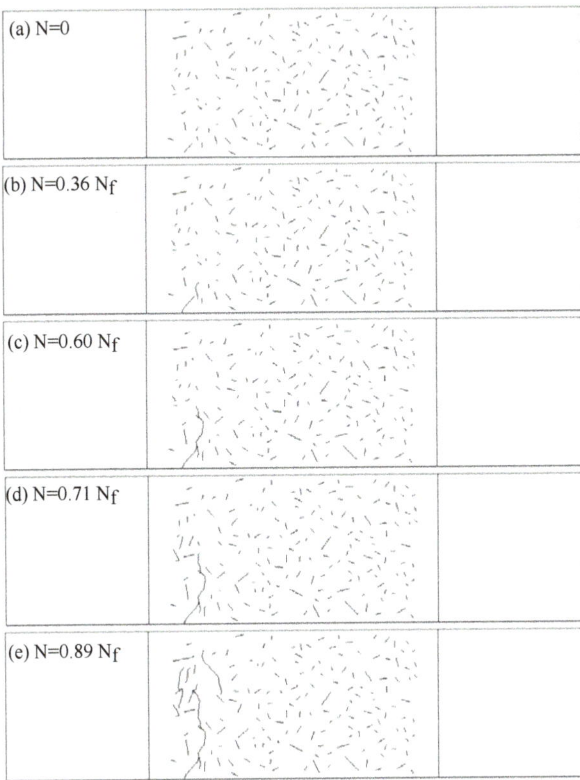

Figure 9. Evolution of crack diagrams.

3.4. Influence of Stress Level on Fatigue Behaviors

Figure 10 shows different failure patterns under various stress levels, where (a) and (b) represent the final crack configurations at low stress level ($S = 0.4$) and high stress level ($S = 0.9$), respectively.

It could be observed that under low fatigue stress level, a relatively smaller number of microcracks propagate during the failure process. Rupture of the beam is mainly caused by one macrocrack, so the damage is highly localized. While in high fatigue stress situation, it could be seen that a relatively larger number of cracks have been propagating, thus macrocracks appear in large quantities. In rupture state, macrocracks scatter over the entire area. Calculated failure modes are basically consistent with the observed ones in literature [35], which could verify the effectiveness of the numerical model. In the experimental observation, macrocracks firstly emerge at certain local areas. Further crack progression will lead to final failure. It shall be emphasized that the experimental observation phenomenologically agrees with our numerical results. Quantitative agreement is limited.

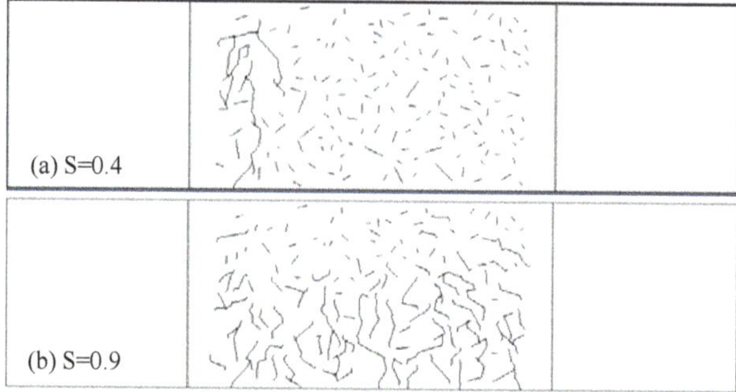

Figure 10. Stress level and failure pattern.

4. Fatigue Damage Analysis of Concrete Beams

Numerical experiments are executed under different fatigue stress levels varying from $S = 0.1$ to $S = 0.9$, from which the fatigue life could be extracted for each stress level. Made possible is the S-N curve that is for engineering reference. Furthermore, since the whole fatigue response has been recorded, macroscopic damage evolution could also be quantitatively analyzed.

4.1. Fatigue Life and S-N Curve

Table 1 lists the calculated results of the numerical experiments. It shows that when the stress level S of the fatigue load decreases from 0.9 to 0.1, the fatigue life of specimen gradually increases from 3000 to 201,000 cycles, refer to Figure 11.

Table 1. Stress level of load and corresponding fatigue life.

No.	S	F_{max} (N)	N_f (10^3)
1	0.9	90	3
2	0.8	80	5
3	0.7	70	10
4	0.6	60	25
5	0.5	50	36
6	0.4	40	42
7	0.3	30	66
8	0.2	20	122
9	0.1	10	201

S-N curve could also be plotted by fitting above data using the classical S-N exponential formula. Note each point represents the fatigue life under designated stress level.

$$N_f = \alpha \exp(-\beta S) \tag{13}$$

Here we have $\alpha = 3.56 \times 10^5$ and $\beta = 5.12$.

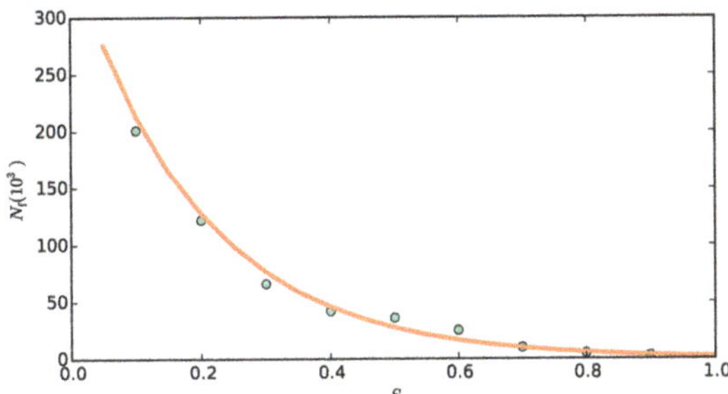

Figure 11. S-N curve obtained by numerical modeling.

4.2. Macroscopic Damage Analysis

Though S-N curve is enough for engineering use to estimate remaining life, it has some limits when it comes to evaluating the current performance state of structure. This is because fatigue life estimation based on S-N curve has an implicit assumption that the fatigue damage is linearly accumulated. In fact, the fatigue damage evolution is a nonlinear process. The period from microcracks propagation and coalescence to the formation of macroscopic cracks is relatively longer, while the speed from the initiation of macrocracks to the final rupture is relatively fast. In this sense, damage evolution based fatigue evaluation is required. According to Lemaitre's concept of damage variable [36], damage is related to stiffness reduction [37,38].

$$D = 1 - \frac{\tilde{K}}{K_0} \tag{14}$$

Here, D is the damage variable, K_0 denotes the effective stiffness in defect-free state, and $\tilde{K}$ is the stiffness in damaged state. Obviously, the larger D grows, especially near 1, the closer material or structure is to the failure state. For three-point bending beams, effective stiffness could be defined as

$$\tilde{K} = \frac{F}{d}. \tag{15}$$

Here, F and d are mid-span force and displacement respectively, which could be extracted from calculated database. In defect-free or damage-free status, d_0 could be obtained as

$$d_0 = \frac{FL^3}{48EI}, \tag{16}$$

where E is the initial Young's modulus of concrete.

Curves of damage evolution at different stress levels are plotted in Figure 12. It can be observed that the fatigue damage evolution of concrete can be roughly divided into three stages. When the cycle ratio n/N_f is less than 0.2, the damage value is relatively small, and the damage evolution rate is also slow. Growth rate of most microcracks fall in zone I. Crack coalescence has not occurred yet. With the increase of load cycles, damage evolution enters into a stable stage that the value of damage increases linearly with the number of load cycles. In this period, microscopic crack growth rate may vary from area to area (some areas in zone II, some still in zone I). Microcracks begin to link with others, gradually forming macroscopic cracks (crack coalescence). When cycle ratio reaches 0.8, structure enters into an unstable crack growth stage. Macroscopic cracks continue to propagate forward, merging small or microcracks along the rupture path. Correspondingly, the value of damage rapidly approaches the value of 1, finally leading to the fracture of beams.

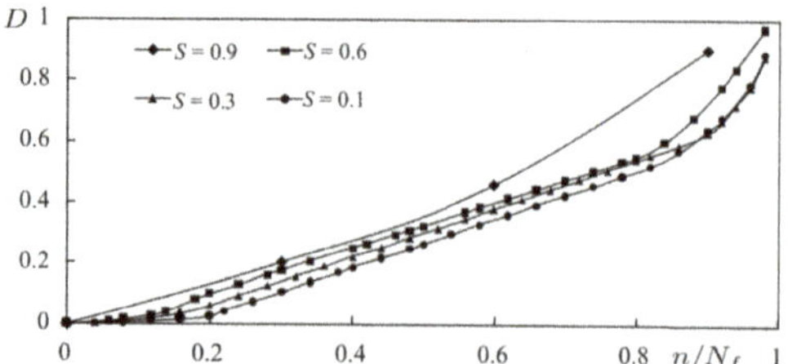

Figure 12. Curves of fatigue damage evolution of concrete beams.

5. Conclusions

The numerical model developed in this paper is directly based on microscopic mechanism of concrete fatigue. Thus it can simulate the fatigue growth of multiple microcracks in concrete, as well as the macroscopic damage evolution. The numerical method is applied to three-point bending beams. Results clearly show that, no matter from macroscopic or microscopic view, fatigue rupture of beams endures three stages: (1) Microcrack growth and coalescence, (2) the formulation of macrocracks and stable fatigue damage evolution, and (3) unstable crack growth and the final fracture of concrete beam.

It is also observed from the simulation that different stress levels of fatigue load can lead to different failure patterns. At low stress level, the rupture of beam is mainly caused by one (or several) macrocrack and damage is highly localized. For high fatigue stress situation, macrocracks appear in large quantities, mostly scattering over the entire area. The computed results are consistent with experimental observation in published literature.

The numerical model could be used to calculate S-N curve. The macroscopic damage evolution of concrete beams is achieved. Based on the results of damage evolution, structural fatigue assessment is also made possible.

Although the present work is about plain concrete, the method can also be applied to other types of concretes, such as high-performance concrete or fiber reinforce concrete, or even masonry materials and structures, as long as the fatigue failure is mainly caused by the growth of microcracks. Last but not least, it shall be pointed out that the numerical modeling is currently stipulated to 2D problems. Computational cost will be substantially added once it comes to 3D. Computational efficiency shall be further improved in our forthcoming work.

Author Contributions: Conceptualization, B.W., K.T. and Z.L.; methodology, code and software, B.W.; data analysis, K.W.; writing—original draft preparation, B.W.; writing—review and editing, K.T. and B.W.; supervision, Z.L.; funding acquisition, Z.L. and K.T.

Funding: This research was funded by National Natural Science Foundation of China under grant No. 11872278 and 51878154, and by the National Program on Major Research Project under grant No. 2016YFC0701301.

Conflicts of Interest: The authors declare no conflict of interest.

References

1. Ovuoba, B.; Prinz, G.S. Investigation of residual fatigue life in shear studs of existing composite bridge girders following decades of traffic loading. *Eng. Struct.* **2018**, *161*, 134–145. [CrossRef]
2. Dias, I.F.; Oliver, J.; Lemos, J.V.; Lloberas-Valls, O. Modeling tensile crack propagation in concrete gravity dams via crack-path-field and strain injection techniques. *Eng. Fract. Mech.* **2016**, *154*, 288–310. [CrossRef]
3. Sun, Z.H.; Scherer, G.W. Effect of air voids on salt scaling and internal freezing. *Cem. Concr. Res.* **2010**, *40*, 260–270. [CrossRef]
4. D'Amato, M.; Laterza, M.; Casamassima, V.M. Seismic performance evaluation of a multi-span existing masonry arch bridge. *Open Civ. Eng. J.* **2017**, *11*, 1191–1207. [CrossRef]
5. Laterza, M.; D'Amato, M.; Braga, F.; Gigliotti, R. Extension to rectangular section of an analytical model for concrete confined by steel stirrups and/or FRP jackets. *Compos. Struct.* **2017**, *176*, 910–922. [CrossRef]

6.   Mander, J.B.; Priestley, M.J.N.; Park, R. Theoretical Stress-Strain Model for Confined Concrete. *J. Struct. Eng.-ASCE* **1988**, *114*, 1804–1826. [CrossRef]

7.   Khalilpour, S.; BaniAsad, E.; Dehestani, M. A review on concrete fracture energy and effective parameters. *Cem. Concr. Res.* **2019**, *120*, 294–321. [CrossRef]

8.   Ding, Z.D.; Li, J. A physically motivated model for fatigue damage of concrete. *Int. J. Damage Mech.* **2018**, *27*, 1192–1212. [CrossRef]

9.   Feng, D.C.; Ren, X.D.; Li, J. Stochastic damage hysteretic model for concrete based on micromechanical approach. *Int. J. Non-Linear Mech.* **2016**, *83*, 15–25. [CrossRef]

10. Mai, S.H.; Le-Corre, F.; Foret, G.; Nedjar, B. A continuum damage modeling of quasi-static fatigue strength of plain concrete. *Int. J. Fatigue* **2012**, *37*, 79–85. [CrossRef]

11. Ray, S.; Kishen, J.M.C. Fatigue crack growth due to overloads in plain concrete using scaling laws. *Sadhana-Acad. Proc. Eng. Sci.* **2012**, *37*, 107–124. [CrossRef]

12. Simon, K.M.; Kishen, J.M.C. A multiscale approach for modeling fatigue crack growth in concrete. *Int. J. Fatigue* **2017**, *98*, 1–13. [CrossRef]

13. Papa, E.; Taliercio, A. Anisotropic damage model for the multiaxial static and fatigue behaviour of plain concrete. *Eng. Fract. Mech.* **1996**, *55*, 163–179. [CrossRef]

14. Rezazadeh, M.; Carvelli, V. A damage model for high-cycle fatigue behavior of bond between FRP bar and concrete. *Int. J. Fatigue* **2018**, *111*, 101–111. [CrossRef]

15. Fan, Z.; Sun, Y. Detecting and evaluation of fatigue damage in concrete with industrial computed tomography technology. *Constr. Build. Mater.* **2019**, *223*, 794–805. [CrossRef]

16. Baktheer, A.; Hegger, J.; Chudoba, R. Enhanced assessment rule for concrete fatigue under compression considering the nonlinear effect of loading sequence. *Int. J. Fatigue* **2019**, *126*, 130–142. [CrossRef]

17. Keerthana, K.; Kishen, J.M.C. An experimental and analytical study on fatigue damage in concrete under variable amplitude loading. *Int. J. Fatigue* **2018**, *111*, 278–288. [CrossRef]

18. Ding, Z.D.; Feng, D.C.; Ren, X.D.; Wang, J.F. Physically based constitutive modeling of concrete fatigue and practical numerical method for cyclic loading simulation. *Eng. Fail. Anal.* **2019**, *101*, 230–242. [CrossRef]

19. Maragoni, L.; Carraro, P.A.; Peron, M.; Quaresimin, M. Fatigue behaviour of glass/epoxy laminates in the presence of voids. *Int. J. Fatigue* **2017**, *95*, 18–28. [CrossRef]

20. Sisodia, S.M.; Gamstedt, E.K.; Edgren, F.; Varna, J. Effects of voids on quasi-static and tension fatigue behaviour of carbon-fibre composite laminates. *J. Compos. Mater.* **2015**, *49*, 2137–2148. [CrossRef]

21. Malekan, M.; Carvalho, H. Analysis of a main fatigue crack interaction with multiple micro-cracks/voids in a compact tension specimen repaired by stop-hole technique. *J. Strain Anal. Eng. Des.* **2018**, *53*, 648–662. [CrossRef]

22. Bazant, Z.P.; Xu, K.M. Size Effect in Fatigue Fracture of Concrete. *ACI Mater. J.* **1991**, *88*, 390–399.

23. Bazant, Z.P.; Schell, W.F. Fatigue Fracture of High-Strength Concrete and Size Effect. *ACI Mater. J.* **1993**, *90*, 472–478.

24. Paggi, M. Modelling fatigue in quasi-brittle materials with incomplete self-similarity concepts. *Mater. Struct.* **2011**, *44*, 659–670. [CrossRef]

25. Bhowmik, S.; Ray, S. An improved crack propagation model for plain concrete under fatigue loading. *Eng. Fract. Mech.* **2018**, *191*, 365–382. [CrossRef]

26. Roe, K.L.; Siegmund, T. An irreversible cohesive zone model for interface fatigue crack growth simulation. *Eng. Fract. Mech.* **2003**, *70*, 209–232. [CrossRef]

27. Del Busto, S.; Betegon, C.; Martinez-Paneda, E. A cohesive zone framework for environmentally assisted fatigue. *Eng. Fract. Mech.* **2017**, *185*, 210–226. [CrossRef]

28. Kirane, K.; Bazant, Z.P. Microplane damage model for fatigue of quasibrittle materials: Sub-critical crack growth, lifetime and residual strength. *Int. J. Fatigue* **2015**, *70*, 93–105. [CrossRef]

29. Nguyen, N.H.T.; Bui, H.H.; Kodikara, J.; Arooran, S.; Darve, F. A discrete element modeling approach for fatigue damage growth in cemented materials. *Int. J. Plast.* **2019**, *112*, 68–88. [CrossRef]

30. Wang, Y.P. Physical stochastic damage model for concrete subjected to fatigue loading. *Int. J. Fatigue* **2019**, *121*, 191–196. [CrossRef]

31. Wu, B.; Tang, K. Modelling on crack propagation behaviours at concrete matrix-aggregate interface. *Fatigue Fract. Eng. Mater. Struct.* **2019**, *42*, 1803–1814. [CrossRef]

Appl. Sci. **2019**, *9*, 4664

32. Wu, B.; Li, Z.; Tang, K. Numerical modeling and analysis on micro-to-macro evolution of crack network for concrete materials. **2019**, submitted.
33. Paris, P.; Erdogan, F. A critical analysis of crack propagation laws. *J. Basic Eng.* **1963**, *85*, 528–533. [CrossRef]
34. Lou, J.; Bhalerao, K.; Soboyejo, A.B.O.; Soboyejo, W.O. An investigation of the effects of mix strength on the fracture and fatigue behavior of concrete mortar. *J. Mater. Sci.* **2006**, *41*, 6973–6977. [CrossRef]
35. Chen, C.; Cheng, L.J. Fatigue Behavior and Prediction of NSM CFRP-Strengthened Reinforced Concrete Beams. *J. Compos. Constr.* **2016**, *20*. [CrossRef]
36. Lemaitre, J. *A Course on Damage Mechanics*; Springer Science & Business Media: Berlin/Heidelberg, Germany, 2010.
37. Tang, K.K.; Li, Z.X.; Wang, J. Numerical simulation of damage evolution in multi-pass wire drawing process and its applications. *Mater. Des.* **2011**, *32*, 3299–3311. [CrossRef]
38. Tang, K.K.; Li, Z.X.; He, D.D.; Zhang, Z.H. Evolution of plastic damage in welded joint of steel truss with pre-existing defects. *Theor. Appl. Fract. Mech.* **2010**, *54*, 117–126. [CrossRef]

© 2019 by the authors. Licensee MDPI, Basel, Switzerland. This article is an open access article distributed under the terms and conditions of the Creative Commons Attribution (CC BY) license (http://creativecommons.org/licenses/by/4.0/).

Article

Assessment of Fatigue Lifetime and Characterization of Fatigue Crack Behavior of Aluminium Scroll Compressor Using C-Specimen

Sang-Youn Park [1], Jungsub Lee [1], Jong-Tae Heo [2], Gyeong Beom Lee [2], Hyun Hui Kim [2] and Byoung-Ho Choi [1,*]

[1]   School of Mechanical Engineering, Korea University, Seoul 02841, Korea; magicparkkor@korea.ac.kr (S.-Y.P.); bouce@korea.ac.kr (J.L.)
[2]   R&D Center, LG Electronic Inc., Seoul 08503, Korea; Hjt.heo@lge.com (J.-T.H.); gyeongbeom.lee@lge.com (G.B.L.); hyunhui.kim@lge.com (H.H.K.)
*   Correspondence: bhchoi@korea.ac.kr; Tel.: +82-2-3290-3378

Received: 31 March 2020; Accepted: 2 May 2020; Published: 6 May 2020

Abstract: Currently, a scroll compressor is used in many industrial fields; hence, research on its reliability is important. The major loading type during operation is pressure. However, unexpected contact between scroll compressors typically occurs, and thus, the severe loading condition should be considered. To consider this condition, the study modified the scroll compressor's structure to a C-type specimen. The study applied cyclic axial load in the specimen. The main objective of the study is to define a proper fatigue life method for the aluminium scroll compressor and proper finite element method (FEM) modeling. To define the method, the study included a case study for various parameters, such as mean stress effects. Furthermore, a crack propagation study is presented. In the study, the Darveaux method that considers the Bauschinger effect of ductile material is used. It is expected that the consideration of the parameters can help define the fatigue life assessment of an aluminium scroll compressor.

Keywords: life assessment; crack initiation; crack propagation; finite element method; scroll compressor

1. Introduction

In recent years, scroll compressors have been widely used in the automobile industry. They are operated under axial and radial loading conditions [1]. The rotation of the scroll leads to pressure and temperature changes. Thus, reliability is an important issue for the scroll. Yaubin Yang [2] attempted to simulate the pressure using ANSYS software. This research studied thermal stress during operation. Furthermore, Christophe Ancel [3] examined fatigue simulation via nCode DesignLife. Christophe used Dang Van analysis to assume the pressure during the operation. The dynamic model was studied by Qiang et al. [4]. He defined the scroll compressor's dynamic motion using dynamic characteristics such as pressure distribution, tangential force, and axial force. Furthermore, the accelerated condition was studied by Chang [5]. This study carried out research about wear under the accelerated condition. By using temperature as a parameter, we can define the crack initiation condition more thoroughly. Various types of gas force in scroll compressors were considered by Wang et al. [6]. By calculating the axial and radical force of gas during cycle, they ran the ANSYS program to calculate the deformation of the scroll. Dominique Gross [7] studied rotation speed. In Gross's study, the structure received axial, tangential and friction forces. For various RPM, they calculated force, which is sometimes more serious for the structure than pressure. After these processes, they predicted fatigue life with the stress–life curve of the material. To reduce the contact force, Lantian Ji [8] applied diamond-like carbon (DLC) film to the compressor. By reducing friction force, resistance can be reinforced. The commonality of the studies [2–7] is that they only calculated regular operation, and did not consider the scroll compressor itself. However, during the operation, the scroll compressor came under an unexpected loading condition, specifically, contact with the counterpart, i.e., the fixed scroll. Although previous research mentioned that the scroll compressor easily succumbs to failure during the operation, there is only a small amount of research which concentrates on the scroll compressor itself. Even though they ran the finite element method (FEM) for the scroll compressor, the validation of the structure was operated under the operating condition, which was dominated by pressure and thermal conditions in those references. Thus, the design of more conservative experiments is needed, which was not studied in previous research. To achieve this requirement, the study modified the scroll compressor to a C-specimen, thereby making it possible to apply concentrated force at the weak point. Furthermore, the fatigue parameter should also be considered with changes in the loading condition. This is because previous research [2,3,5,7] tends to focus on high cycle fatigue under elastic deformation. To recommend a proper fatigue parameter, fatigue life should be divided into two regions: initiation and propagation [9].

Crack initiation follows Miner's rule [10]. The rule states that, although the application of stress may not be sufficient to induce failure, the accumulation of repeating stress can lead to failure. If the damage accumulation arrives at one, then it implies crack initiation. The Stress–life (S-N) curve method is generally applied [9] to determine this cycle. This method is advantageous because measurements can be easily performed. The relationship between amplitude stress and fatigue cycle is given in Equation (1) as follows; [9].

$$\sigma_a = \sigma'_f \times (2N)^b \tag{1}$$

where N is the life cycle, σ_a is amplitude stress, σ'_f is the fatigue strength coefficient and b is fatigue strength exponent.

In addition to the fatigue curve, the mean stress effects must also be considered. The mean stress effect is a counting method for correcting non-uniform loading conditions [9]. Typically, the Goodman method (Equation (2)) and Gerber method (Equation (3)) [9] are used to correct non-uniform loading conditions. The Gerber method is less conservative when compared with the Goodman method. Additionally, in case of brittle materials, both methods tend to be similar to the Goodman method [9]. However, actual test data tend to exist between the Goodman and Gerber lines.

$$\frac{\sigma_a}{\sigma_{ar}} + \frac{\sigma_m}{\sigma_u} = 1 \tag{2}$$

$$\frac{\sigma_a}{\sigma_{ar}} + \left(\frac{\sigma_m}{\sigma_u}\right)^2 = 1 \tag{3}$$

where σ_{ar} is the equivalent fully reversed stress amplitude resulting in the same fatigue life, σ_a is the amplitude stress, σ_m is the mean stress and σ_u is the ultimate stress.

However, the methods tend to be inaccurate in severe loading conditions wherein plastic deformation can occur in the structure. The strain–life (ε-N) curve should be considered to solve the problem [11]. The method is advantageous because it can depict the Bauschinger effect of a material. By highlighting the Bauschinger effect, the method can consider the plastic deformation of the material between loading and unloading regions. The method is given in Equation (4) as follows; [11].

$$\frac{\varepsilon_a}{2} = \frac{\sigma'_f}{E} \times (2N)^b + \varepsilon'_f (2N)^c \tag{4}$$

where ε_a is the strain amplitude, σ'_f is the fatigue strength coefficient, N is life cycle, E is young's Modulus, b is the fatigue strength exponent, ε'_f is the fatigue ductility coefficient and c is the fatigue ductility exponent.

With respect to the strain(ε)–life(N) curve, there are two mean stress effect correction models, the Morrow (Equation (5)) [12] and Smith Watson Topper (SWT) (Equation (6)) models [13]. Ince et al. [14] indicated that the SWT model can apply changes in the load by calculating the energy that is multiplied with σ_{max}. This method is rather flexible, especially in high loading conditions.

$$\frac{\varepsilon_a}{2} = \frac{\sigma'_f - \sigma_m}{E}(2N)^b + \varepsilon'_f\left(\frac{\sigma'_f - \sigma_m}{\sigma'_f}\right)^{\frac{c}{b}}(2N)^c \tag{5}$$

$$\sigma_{max}\varepsilon_a = \frac{\left(\sigma'_f\right)^2}{E}(2N)^b + \varepsilon'_f\sigma'_f(2N)^{b+c} \tag{6}$$

where σ_{max} is the max stress during the cycle.

Based on these mean stress effects, many researchers developed models. Ince [15] suggested a multiaxial fatigue damage model based on strain energy. By considering strain shear energy, multi axial fatigue life can be considered. This method can also be adopted in distortion [16]. The form of energy was similar with that of SWT, which can be calculated by multiplying the strain and force of each type. There is little research applying Optistruct in the scroll compressor. However, there are another cases using Optistruct in fatigue simulation. Kim [17] calculated the fatigue life of the skate frame using FEM. In this study, he studied the surface effect and size effect, which can modify the fatigue limit. Musaddiq [18] compared the stress–life and strain–life curve in calculating the fatigue life of the joint prosthesis using FEM. He suggested that, for composite structures which receive excessive force, the strain–life curve is a proper parameter rather than the stress–life curve.

The fatigue lifetime of a structure can be divided into three mechanisms, i.e., crack initiation, crack propagation and dynamic instability. Among them, the time for dynamic instability is extremely short compared with the other two mechanisms. However, the contribution of crack propagation on the entire lifetime of a structure should be considered, unless the material fabricating a structure is brittle enough. Therefore, the lifetime of the scroll compressor should be studied by considering both the crack initiation and crack propagation behaviors. Hence, crack propagation should be considered to predict the lifetime until failure of the scroll compressor. The general method to depict crack propagation is the Paris law as follows; [19,20].

$$\frac{da}{dN} = \Delta G > G_{1c} \tag{7}$$

where N is the life cycle, a is the length of the crack, G is the energy release rate, G_{1c} is the critical energy release rate.

This equation presents that if the energy release rate exceeds critical energy release rate, the crack propagates. This method is based on the linear elastic fracture mechanism (LFEM). Thus, if plastic deformation occurs in the structure, it tends to be inaccurate. To compensate for the weakness, the Darveaux method [20,21] should be considered (Equation (8) and Figure 1). By using strain energy density per cycle, the Darveaux coefficient is calculated as follows; [20,21].

$$N_0 = c_1(\Delta w)^{c_2},\ D_{N+\Delta N} = D_N + \frac{\Delta N}{L}c_3(\Delta w)^{c_4}\ (c_1, c_2, c_3, c_4 - material\ constant) \tag{8}$$

where N_0 is cycle number for damage initiation, Δw is inelastic hysteresis strain criteria. D_N is damage variable for degradation stiffness assumption. Thus, the degradation of the element can be calculated as follows; [20].

$$\sigma = (1 - D)\overline{\sigma},\ \text{is the stress of the current increment.} \tag{9}$$

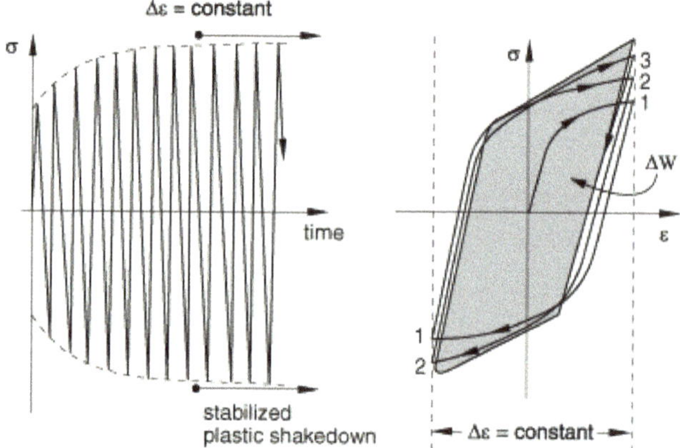

Figure 1. Plastic shakedown in a direct cyclic analysis [20].

The Darveuax method was used in many recent studies. Liulu [22] used the Darveux method to predict solder joints. By tracking individual solder failures, he mentioned that the Darveuax method can depict the change of inelastic region during repeated cyclic load. Thus, he suggested that the Darveuax method have advantages for depicting ductile material due to the inelastic strain energy density accumulated per cycle. Liuyang [23] proposed a method to calibrate the damage material parameters using both the Darveuax method and the strain–life curve. Even though there was no recent research which applied the Darveuax method to the scroll compressor, these recent studies have become helpful examples.

In the study, a parametric investigation is presented to determine the appropriate parameter for the customized C-specimen test. The entire cycle is calculated separately by dividing the initiation and propagation cycles. The crack initiation cycle is calculated by using Altair Optistruct software, which can accumulate element damage during cyclic loading. Specifically, ABAQUS 6.14 is used to calculate the crack propagation cycle. The specific results of the analysis are expected to aid in determining safe specifications of the scroll compressor.

2. Experiments

The scroll compressor used in this study is made of a silicon-manganese aluminium alloy, i.e., Al 4007A, as it is quite difficult to evaluate fatigue characteristics in the scroll compressor due to its complicated geometry and loading conditions. Therefore, a newly designed specimen, by considering the most dangerous region under operating conditions, is proposed and used in this study. To design the customized C-specimen, the stress distribution and deformation profile during the operation are considered, as shown in Figure 2. Prior to performing the experiment with the C-specimen, 4007A's material fatigue curve (S-N curve) was evaluated. A rotary bending test machine from Shimadzu Corporation, Kyoto, Japan was utilized. The advantage of the machine is that it increases speed to 50 Hz, which is five times faster than a tensile test machine. Furthermore, the rotary bending machine's fatigue R-ratio corresponds to −1, which is the ideal number for the fatigue curve.

As previously mentioned, the structure of C-specimen was revised based on the operating conditions, as shown in Figure 2, to apply severe loads comparable with the pressure. By removing a part of the wrap, the experimental device can accurately apply a vertical load at Point A, which is the weak point during scroll compressor operation as per operation and simulation data. The experimental device for the fatigue test of C-specimen was MTS 810 (MTS Systems Corp., Minneapolis, MN, USA). An additional guide structure was designed and used to prevent unexpected loading types (e.g., tangential loading between zig and wrap) as shown in Figure 3. The operation frequency was 10 Hz, and the R-ratio of the loading condition was 0.05. Additionally, five amplitude loads were applied as follows: 1900, 1662.5, 1444, 1140, and 948 N.

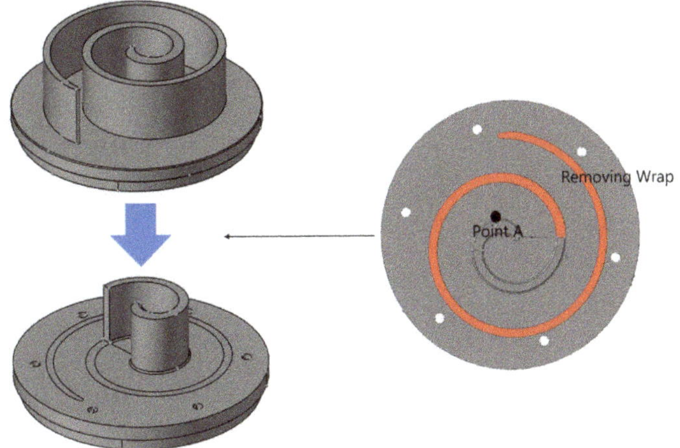

Figure 2. Modified scroll compressor to C-specimen.

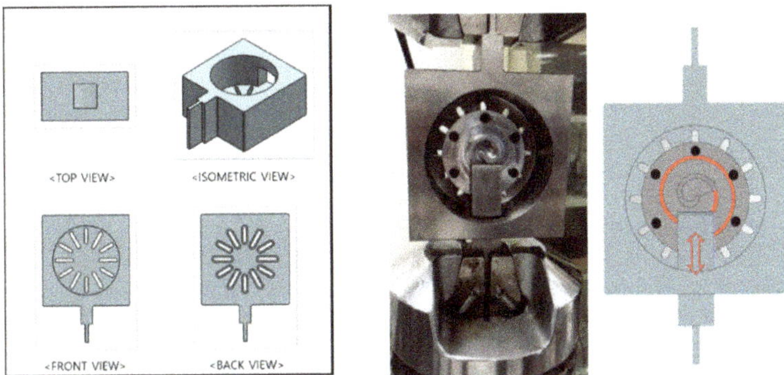

Figure 3. Guide structure (left) and experiment setting (right).

3. Finite Element Modeling Information

3.1. Material Information

Aluminium 4007A was used as the material for the scroll compressor [24]. The composition of the material is described in Table 1. The 4000 series aluminium contains more than 1% silicon, and thus the material is generally characterized with high yield strength and low elongation when compared with that of other aluminium materials.

Table 1. Composition of the aluminium alloy 4007A [24] (unit: %).

Elements	Al	Si	Mn	Fe	Ni	Cr
Percentage	96.3	1.4	1.2	0.7	0.3	0.1

The mechanical properties of the tensile test, listed in Table 2, were used in Optistruct and ABAQUS. In order to use Optistruct, stress–life and strain–life curves should be used. The parameters used in Optistruct are listed in Table 3. Essentially, fatigue strength coefficient (σ'_f) and exponent (b) are identical in the S-N and ε-N curves. They can be calculated from the S-N curve (Figure 4) measured from the rotary bending machine. The equation used for the other parameters is calculated from [9].

Table 2. Mechanical properties of Al 4007A.

Elements	Young's Modulus	Yield Strength	Tensile Strength	Elongation to Break
Percentage	70.0 GPa	340 Mpa	405 Mpa	0.09

Table 3. Strain–life curve information in Optistruct.

Symbol	σ'_f	ε'_f	b	C	K'	n'
Value	723 Mpa	0.11	−0.091	−0.705	455 Mpa	0.12
Equation	-	Elongation	-	$n' = \frac{c}{b}$ [9]	From S-S curve [9]	From S-S curve [9]

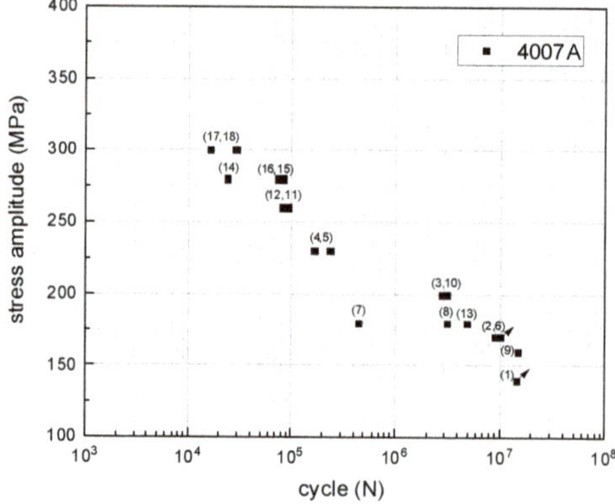

Figure 4. Stress–Life curve of Al 4007A.

With respect to the Paris law and Darveaux method, the following parameters in Table 4 are used. If the model only considers Paris law, G_{1c} is used. To consider the Darveaux method, $c_1 \sim c_4$ should be added. The parameters were calculated from the cyclic S-S curve of the material from the tensile test machine.

Table 4. Values used in the Paris law and Darveux coefficient in ABAQUS extend finite element method (XFEM).

Symbol	G_{1c}	c_1 (Cycles/MPa^{c_2})	c_2	c_3	c_4
Value	0.014 Mpa	0.9×10^{-7}	−0.1	6×10^{-5}	0.327

3.2. FEM Modeling

3.2.1. Loading and Boundary Condition

Geometry and boundary conditions are identical for simulations in Optistruct and ABAQUS. The size of the element was 0.5 mm in Optistruct and ABAQUS. The type of the element was a hexagon. The force acting on the nodes was used to depict the loading condition. The total amount of loads was summary of the load given at each node. The red area shown in Figure 5 is the fixed area in FEM. This is because if the guide structure is used, the complete back side of C-specimen would be fixed firmly, even though only six bolting points were used in the actual experiment. This is the same effect as the structure fixed on the wall. Thus, all degree of freedom (DOF) of nodes on the back side were fixed. In the study, crack initiation and crack propagation were measured separately. Total cycle can be counted via a summary of cycle results of Optistruct and ABAQUS XFEM.

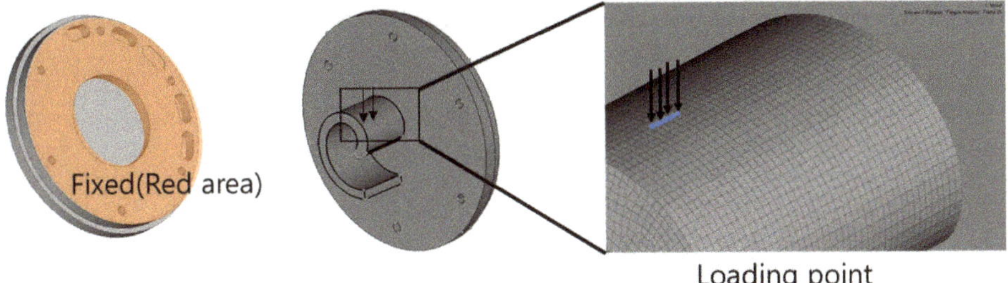

Figure 5. Boundary and loading conditions of the finite element method (FEM).

3.2.2. Crack Initiation

With respect to crack initiation, the study used Altair Optistruct 2018. If the crack occurred, it represented the damage contour as 1 (Figure 6) based on the Miner's rule [10]. Both Optistruct and ABAQUS software run fatigue based on one static step. This means that one cycle step should be utilized as an initial condition before using the fatigue step to run the fatigue module. The study used both stress–life curve and strain–life curve in Optistruct. Thus, by performing the parametric study for various mean stress effects such as the Goodman, the Gerber, the Morrow and the SWT, the study recommends proper parameters by comparing the results of the experiment and FEM.

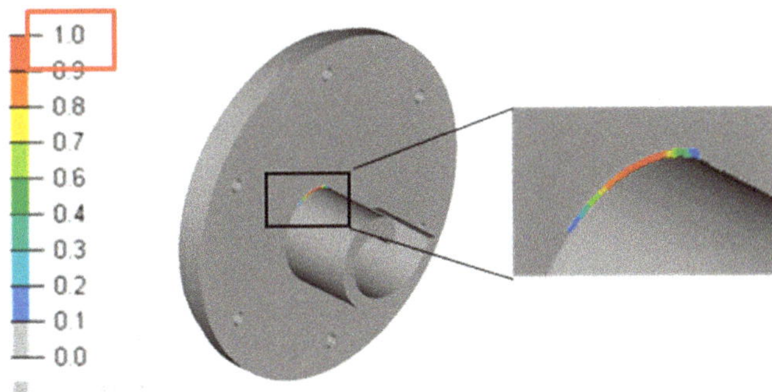

Figure 6. Damage contour of fatigue simulation in Optistruct.

3.2.3. Crack Propagation

Crack propagation was examined via ABAQUS extend finite element method (XFEM). When ABAQUS XFEM is used, it is assumed that crack is already initiated, and that the location of crack initiation is accurate. Specifically, ABAQUS XFEM depicts the crack propagation line that connects between virtual nodes, which are not on the element. The advantage of the method is that it conserves the stiffness of the structure because it does not delete the element. The basic enrichment function of the XFEM is in Figure 7 and Equations (10)–(12) [25].

$$u = \sum_{I=1}^{N} N_I(x)\left[u_I + H(x)a_I + \sum_{\alpha=1}^{4} F_\alpha(x)b_I^\alpha\right] \tag{10}$$

$$H(x) \begin{cases} 1, & if\ (x - x^*) * n \geq 0 \\ -1, & otherwise \end{cases} \quad x^* \text{ is the point on the crack closet to x} \tag{11}$$

$$F_\alpha(x) = \left[\sqrt{r}sin\frac{\theta}{2},\ \sqrt{r}cos\frac{\theta}{2},\ \sqrt{r}sin\theta sin\frac{\theta}{2}, sin\theta cos\frac{\theta}{2} \right], \tag{12}$$

where $N_1(x)$ is the usual nodal shape functions and u_I is the displacement vector of the elements in the part. $H(x)$ is the discontinuous function which can be presented like equation 11 [25]. $F_\alpha(x)$ is the stress asymptotic function of crack tip can be represented as Equation (12) [25]. (r, θ) in Equation (12), is polar coordinate from origin crack tip. a_I and b_I^α in Equation (10), are nodal enriched degrees of freedom vectors of $H(x)$ and $F_\alpha(x)$.

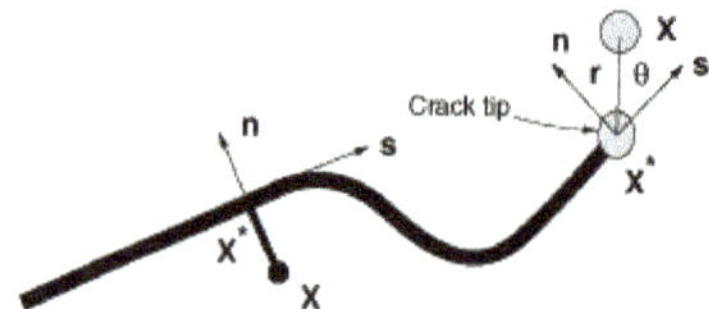

Figure 7. Illustration of normal and tangential coordinates for a smooth crack [25].

Equation (10) consists of the near tip asymptotic functions that capture singularity around the crack tip and a discontinuity that represents the jump in displacement across the crack surfaces. u_I is applicable for all nodes. $H(x)a_I$ is cut by crack. $\sum_{\alpha=1}^{4} F_\alpha(x)b_I^\alpha$ is used only for nodes whose shape function is cut by the crack tip.

To decide the direction parameter, (r, θ), the XFEM crack looks for the location which satisfies damage initiation criterion. Without this criterion, the crack will just extend without a crack path change (θ = 0). If the Paris law is used without the Darveuax coefficient, the onset of delamination will follow Equation (7), which satisfies $\Delta G > G_{1c}$. However, if the damage initiation criterion is used, such as the Darveuax method, it will change the damage variable of the element, D, used in Equation (8). This means that ΔG changes due to degradation of the element. This degradation can change the crack path by changing stress distribution near the crack tip based on Equation (9).

To count the cycle, the direct cycling step in ABAQUS was used. Based on the results of the static step, the program calculates the cycle to propagate. The important assumption made in order to run ABAQUS XFEM is that the location of the initial crack is exact. In the study, the crack initiation point is same with the location of Optistruct results. Thus, the crack initiation point is based on the results of Optistruct as presented in Figure 6. Constants for the empirical Paris law are considered in XFEM. To apply the Darveux coefficient, a material keyword, * Damage initiation/evolution, should be used. A basic elastic–plastic curve and boundary condition are identical to those of Optistruct. The effect of the Darveuax method is presented in the next chapter, by comparing the model which used the Darveuax method in its material keyword and the model that did not. If the Darveux method is not used, the crack will be only be propagated with the fracture criterion keyword using the critical energy release rate without strain energy change.

4. Results

4.1. Experiment

Figure 8 shows an example of the load–displacement curve of the C-specimen. The displacement tends to rise significantly when the crack initiates and propagates. This is because after the crack initiates, the wrap tends to bend more than before.

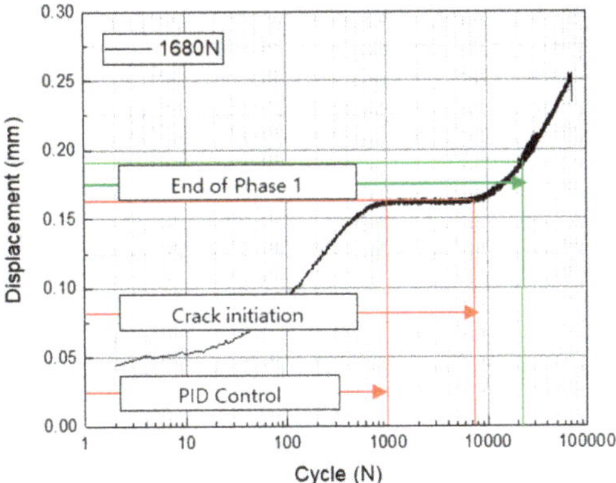

Figure 8. Example of displacement–cycle curve for the c-specimen at 1680 N.

The typical shape of the crack is shown in Figure 9. The crack initiated at the root below the loading point (Figure 9). After a crack initiates, the crack propagates along the wrap's root shape (Phase 1 in Figures 8 and 9). After a certain number of cycles from the end of Phase 1, the crack changes direction outside from the wrap root (Phase 2 in Figure 10). In this phase, the displacement is approximately about 1.2 times larger than the displacement of crack initiation. The fracture mode implies fracture of the wrap. The length of Phase 1 tends to increase when the applied load decreases. This is because a decrease in the load delays severe fracture and Phase 2 in the wrap.

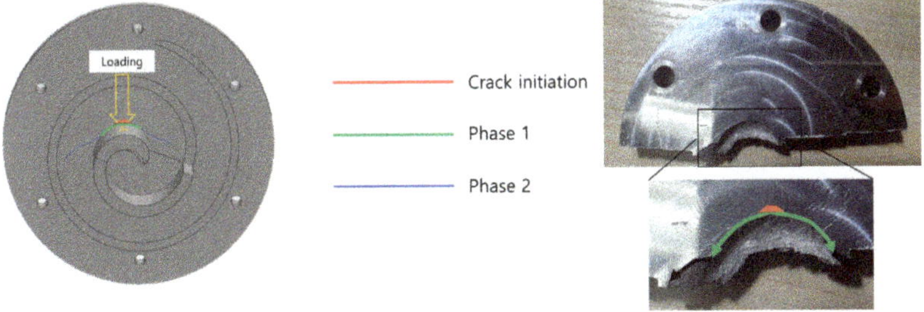

Figure 9. Crack shape picture and example of crack occurrence when the amplitude load is 1920 N.

Figure 10 shows the load amplitude and cycle curve of crack initiation, namely Phase 1. The curve shape of the crack initiation and Phase 1 is similar. Hence, it is estimated that all phases are affected by the same mechanical behavior. For example, in the case of the load exceeding 1900 N, at which plastic deformation occurs, all phases in progress should consider the Bauschinger effect.

4.2. Finite Element Method

4.2.1. Crack Initiation

In this study, a parametric investigation with respect to two parameters is performed. The first parameter is criterion of damage. We recommend using max principal stress as a criterion. This is because if the von Mises criterion is applied, then the damage contour represents the crack initiation point at the maximum compression location as opposed to the tensile maximum point (Figure 11). However, in the experiment, no crack is observed at this location. If the max principal stress is selected as the criterion, it identifies the accurate location, which is similar to the results of the experiment.

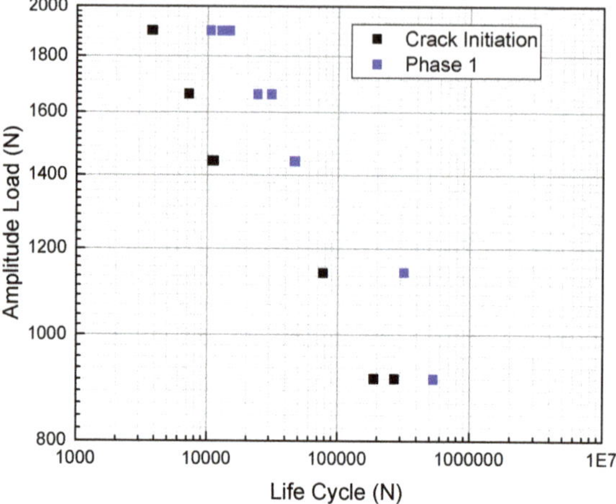

Figure 10. Load amplitude–Cycle curve of the C-specimen.

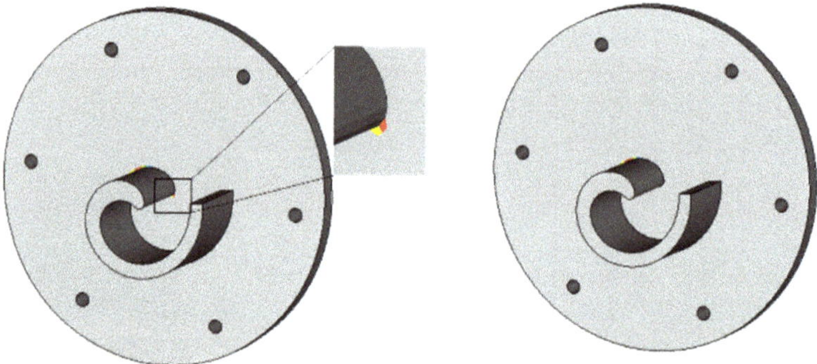

Figure 11. Damage contour of fatigue FEM for various criteria: von Mises (left)/max principal stress (right).

With respect to stress–life curve, the Gerber and Goodman methods were examined. Figure 12 shows the FEM results of the fatigue analysis when the Goodman and Gerber methods were used. As mentioned in the previous chapter, the Goodman method is a conservative method that exhibits a lower cycle than the Gerber method. Thus, when the applied amplitude load exceeds 1500 N, the results of the Goodman method are relatively inaccurate when compared with that of the Gerber method. If the applied load is sufficiently low, then the experimental results are between those of the Goodman and Gerber methods.

With respect to strain–life curve, the Morrow and SWT models were compared. Figure 13 shows the amplitude of the load and life cycle curve of the structure. Under a load of 1500 N, there is no significant difference between the Morrow and SWT. However, increases in the load increases the difference between the two. This implies that SWT can correct high loading effects by calculating the energy of the structure. It is expected that the SWT can model the experiment more accurately than the Morrow as the load in the experiment increases. The result is similar to that obtained in a study by Dowling [26], which indicates that SWT is more suitable for the aluminium alloy as opposed to the Morrow.

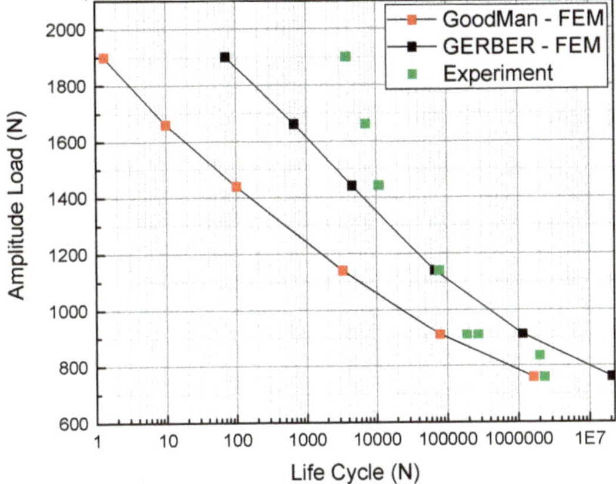

Figure 12. Comparison of the FEM and experiment when the stress–life (S-N) curve is applied in FEM.

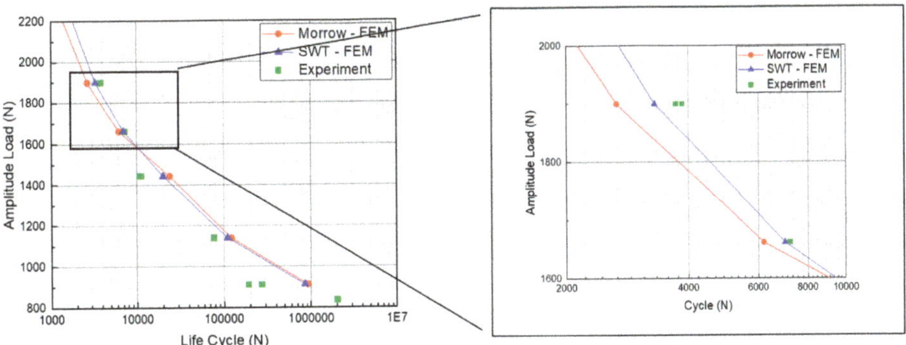

Figure 13. Comparison between FEM and experiment when ε-N curve is applied in FEM.

4.2.2. Crack Propagation

In this subsection, crack propagation models that use the Paris law and hysteresis energy method are compared and validated. Figure 14 shows the crack propagation characteristics predicted by the ABAQUS XFEM when the applied load is varied. It is observed that, in the loading condition under a load amplitude of 1400 N, the difference as to whether the Darveaux method is used is not significant (e.g., Figure 14a with the load amplitude of 916N). This is because plastic deformation does not dominate the structure in the loading condition due to the low loading amplitude. However, if the load amplitude exceeds 1900 N, then the difference is based on the usage of the Darveaux method. As shown in Figure 14b, the direction of the crack is slightly upward when the Darveaux method is not used. This is because the structure cannot apply the Bauschinger effect during crack propagation if the Darveaux method cannot be applied in the structure, and thus the stiffness of the structure is weak when compared with the model with the application of Darveaux method. This assumption can also be proven by the displacement–cycle curve as shown in Figure 15. In the absence of the Darveaux method, the FEM analysis does not reflect the Bauschinger effect. Thus, the speed of crack propagation exceeds that of the experiment and FEM model, which adds the Darveaux coefficient.

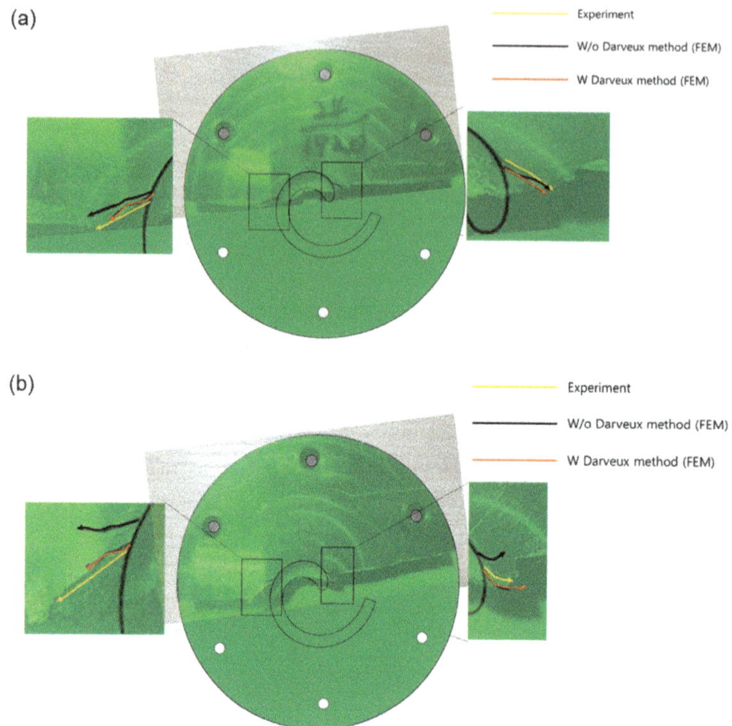

Figure 14. Comparison of the crack path between experiment and FEM. (**a**) Amplitude load of 916N. (**b**) Amplitude load of 1920N.

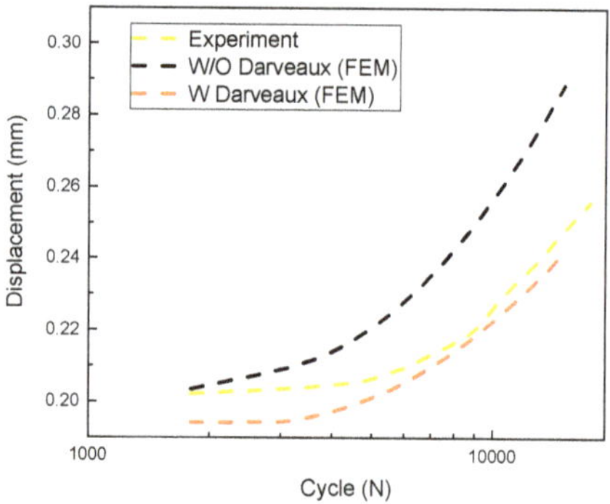

Figure 15. Displacement comparison between FEM and experiment from crack initiation to end of Phase 1 (before this cycle, it is assumed that cycle before crack initiation is correct).

5. Conclusions

In the study, a conservative method for scroll compressor designs was used. The study revised the structure of the scroll compressor by removing part of the wrap. This modification makes zig contact possible on the weak point directly. The life cycle of the scroll compressor was measured continuously after crack initiation. The change of the crack path can be estimated in the displacement-cycle curve of the experiment from the change of the slope. The study divides this region as Phase 1, where the crack changes direction outside from the wrap root. Additionally, various parameters were examined via the FEM. A parametric study was included to determine the correct parameters, such as mean stress effects and crack propagation model.

With respect to the crack initiation cycle, the mean stress effect was examined. For the S-N curve, the Goodman and Gerber method were compared. In a relatively high loading condition, the Gerber method records more cycles than the Goodman method. With a decrease in the loading condition, the experimental results were between the results of the Goodman and Gerber methods. Thus, the Gerber method is recommended while using the S-N curve for the C-specimen fabricated from 4007A. With respect to the ε-N curve model, the Morrow and SWT were compared. In the high loading condition over 1900 N, the SWT tends to be more accurate than Morrow. The energy calculation indicates that the SWT can more efficiently depict the high loading cycle of an aluminium structure.

With respect to crack propagation, the effect of the Darveaux method was examined. In a relatively low loading cycle, there is no significant difference as to whether or not the Darveaux method is used. This is because plastic deformation does not dominate during the fatigue cycle in the conditions. However, in the loading condition over 1900 N, the results differed based on the usage of the Draveaux method. Given its plastic deformation, the model that did not use the Draveaux method tended to crack and propagate upward. The model that used the Draveaux method tended to be more similar than the other model.

In the study, the behavior of the C-specimen is evidently dominated by the Bauschinger effect. To depict the fatigue result, the use of SWT and the Draveaux method are recommended. Thus, it is expected that conservative design can be obtained by using the results of the study.

Author Contributions: Project administration, B.-H.C.; Software, S.-Y.P.; Supervision, J.-T.H.; Validation, J.L., G.B.L. and H.H.K.; Writing–original draft, S.-Y.P. All authors have read and agreed to the published version of the manuscript.

Funding: The research was funded by LG Electronics corporations Inc.

Acknowledgments: The author would like to acknowledge the sharing the scroll compressor and the specimen of LG Electronics Inc.

Conflicts of Interest: The authors declare no conflict of interest.

Nomenclature

σ_a	Strength amplitude
σ_m	Mean stress
σ_u	Ultimate tensile strength
N	Loading cycle
ε_a	Strain Amplitude
σ'_f	Fatigue strength coefficient
b	Fatigue strength exponent
c	Fatigue ductility exponent
ε'_f	Fatigue ductility coefficient
K'	Cyclic strength coefficient
n'	Cyclic strain hardening exponent
D	Damage variable of the element
E	Modulus of elasticity
G	Energy release rate
w	Strain energy density
L	Characteristic length with respect to integration point

References

1. Morishita, E.; Sugihara, M.; Inaba, T.; Nakamura, T. Scroll compressor analytical model. In Proceedings of the International Compressor Engineering Conference, West Lafayette, IN, USA, 11–13 July 1984.
2. Yang, Y.; Tang, Y.-J.; Chang, Y.-C. Static and dynamic analysis on R410A scroll compressor components. In Proceedings of the International Compressor Engineering Conference, West Lafayette, IN, USA, 12–15 July 2010.

3. Ancel, C.; Gross, D.; Guglielmi, L. Fatigue design for scroll compressor wraps. In Proceedings of the International Compressor Engineering Conference, West Lafayette, IN, USA, 12–15 July 2010.

4. Qiang, J.; Peng, B.; Liu, Z. Dynamic model for the orbiting scroll based on the pressures in scroll chambers–Part I: Analytical modeling. *Int. J. Refrig.* **2013**, *36*, 1830–1849. [CrossRef]

5. Chang, M.S.; Park, J.W.; Choi, Y.M.; Park, T.K.; Choi, B.O.; Shin, C.J. Reliability evaluation of scroll compressor for system air conditioner. *J. Mech. Sci. Technol.* **2016**, *30*, 4459–4463. [CrossRef]

6. Wang, H.; Tian, J.; Du, Y.; Hou, X. Numerical simulation of CO_2 scroll compressor in transcritical compression cycle. *Heat Mass Transf.* **2018**, *54*, 1395–1403. [CrossRef]

7. Gross, D.; Genevois, D. Scroll Wrap Additional Stress In Variable Speed Application. In Proceedings of the International Compressor Engineering Conference, West Lafayette, IN, USA, 9–12 July 2018.

8. Ji, L.; He, Z.; Han, Y.; Chen, W.; Xing, Z. Investigation on the Performance Improvement of the Scroll Compressor by DLC Film. In Proceedings of the IOP Conference Series: Materials Science and Engineering, University of London, UK, 9–11 September 2019.

9. Aliabadi, M. *Fundamentals of Metal Fatigue Analysis*; Bannantine, J.A., Comer, J.J., Handrock, J.L., Eds.; Prentice Hall Publishers: Upper Saddle River, NJ, USA, 1990; pp. 2–27.

10. Miner, M. Cumulative fatigue damage. *J. Appl. Mech.* **1945**, *12*, A159–A164.

11. Aliabadi, M. *Fundamentals of Metal Fatigue Analysis*; Bannantine, J.A., Comer, J.J., Handrock, J.L., Eds.; Prentice Hall Publishers: Upper Saddle River, NJ, USA, 1990; pp. 40–72.

12. Morrow, J. Fatigue properties of metals. In *Fatigue Design Handbook*; Society of Automotive Engineers: Pittsburgh, PA, USA, 1968; pp. 21–30.

13. Smith, K. A stress-strain function for the fatigue of metals. *J. Mater.* **1970**, *5*, 767–778.

14. Ince, A.; Glinka, G. A modification of Morrow and Smith–Watson–Topper mean stress correction models. *Fatigue Fract. Eng. Mater. Struct.* **2011**, *34*, 854–867. [CrossRef]

15. Ince, A.; Glinka, G. A generalized fatigue damage parameter for multiaxial fatigue life prediction under proportional and non-proportional loadings. *Int. J. Fatigue* **2014**, *62*, 34–41. [CrossRef]

16. Ince, A. A mean stress correction model for tensile and compressive mean stress fatigue loadings. *Fatigue Fract. Eng. Mater. Struct.* **2017**, *40*, 939–948. [CrossRef]

17. Kim, H.K. Evaluation of Fatigue Endurance Foran Ultra Lightweight Inline Skate Frame. In *Advanced Materials Research*; Trans Tech Publications Ltd: Bäch, Switzerland, 2013.

18. Al-Ali, M.A.; Al-Ali, M.A.; Takezawa, A.; Kitamura, M. Topology Optimization and Fatigue Analysis of Temporomandibular Joint Prosthesis. *World J. Mech.* **2017**, *7*, 323–339. [CrossRef]

19. Paris, P.C. A rational analytic theory of fatigue. *Trend Eng.* **1961**, *13*, 9.

20. Abaqus. ABAQUS user's manual. In *Version 6.14 User's Guide*; Abaqus: Providence, RI, USA, 2014; p. Section 23.2.2.

21. Darveaux, R. Effect of simulation methodology on solder joint crack growth correlation and fatigue life prediction. *J. Electron. Packag.* **2002**, *124*, 147–154. [CrossRef]

22. Jiang, L.; Zhu, W.; He, H. Comparison of darveaux model and coffin-manson model for fatigue life prediction of bga solder joints. In Proceedings of the 2017 18th International Conference on Electronic Packaging Technology (ICEPT), Harbin, China, 16–19 August 2017.

23. Feng, L.; Qian, X. Low cycle fatigue test and enhanced lifetime estimation of high-strength steel S550 under different strain ratios. *Mar. Struct.* **2018**, *61*, 343–360. [CrossRef]

24. Association, A. *Aluminum Standards and Data*; Aluminum Association: Washington, DC, USA, 2013.

25. Abaqus. ABAQUS user's manual. In *Version 6.14 User's Guide*; Abaqus: Providence, RI, USA, 2014; p. Section 10.7.1.

26. Dowling, N.E. *Mean Stress Effects in Stress-Life and Strain-Life Fatigue*; SAE Technical Paper: Pittsburgh, PA, USA, 2004; p. 13.

© 2020 by the authors. Licensee MDPI, Basel, Switzerland. This article is an open access article distributed under the terms and conditions of the Creative Commons Attribution (CC BY) license (http://creativecommons.org/licenses/by/4.0/).

Article

Experimental Investigations on the Effects of Fatigue Crack in Urban Metro Welded Bogie Frame

Wenjing Wang [1,*], Jinyi Bai [1], Shengchuan Wu [2], Jing Zheng [3] and Pingyu Zhou [3]

[1] School of Mechanical, Electronic and Control Engineering, Beijing Jiaotong University, Beijing 100044, China; 17116351@bjtu.edu.cn
[2] State Key Laboratory of Traction Power, Southwest Jiaotong University, Chengdu 610031, China; wusc@swjtu.edu.cn
[3] CRRC Qing Dao SiFang Co., Ltd., Qingdao 266111, China; zhengjing@cqsf.com (J.Z.); zhoupingyu@cqsf.com (P.Y.Z.)
* Correspondence: wjwang@bjtu.edu.cn; Tel.: +86-10-516-831-95

Received: 5 February 2020; Accepted: 20 February 2020; Published: 24 February 2020

Abstract: The welded bogie frame is the critical safety part of the urban metro vehicle. This paper focuses on finding out the factors inducing the fatigue cracks initiated from the positioning block weld toe of metro bogie frame. Fracture morphology and metallographic analysis were conducted to identify the failure modes, and on-track tests about the dynamic stress at the positioning block weld toe and vibration acceleration were performed. The typical signals of dynamic stress and acceleration were analyzed from time and frequency domain. The relationship between wheel polygon, rail corrugation, running speed and dynamic stress in amplitude and frequency are investigated in details. Research results show that the micro cracks induced by welding at the weld toe of positioning block propagate to the spring sleeve under relatively high alternating dynamic stress, which is strongly influenced by the wheel polygon, rail corrugation and the train running speed.

Keywords: fatigue crack; welded bogie frame; wheel polygon; rail corrugation; running speed

---

1. Introduction

With the rapid urbanization and industrialization of China society, a huge metro rail network has been building and running in recent years. By the end of 2019, the operating mileage of China urban rail transit has reached about 6600 km and the vehicles in service have reached more than 35,000. As the most important load-carrying structure for railway vehicles, the welded bogie frame undergoes complex fatigue loads [1,2]. Under such alternating loads, the safety critical zone inside the frame necessarily experiences cumulated fatigue damage which seriously threatens the operation safety of rail passenger vehicles [3–5].

As frequently observed in metro vehicle frames, fatigue cracks are usually initiated from the welded region of the bogie frame due to complex operation loads. Recently, a fatigue crack in the Shinkansen train bogie frame [6] was found with a total length of 44 mm, which is regarded as a serious incident in the Japan Railway system, which could possibly have led to a derailment. Compared with high-speed railway passenger trains with covering super long distances and time, the urban metro vehicles are used to experiencing more abnormal loading due to frequent start-stops and huge passenger uncertainty. Therefore, lots of failures have come out in recent years, for example fatigue cracks observed near the motor suspension seat, beam and positioning block, etc. [7,8].

The initiation mechanism of fatigue cracks has been investigated extensively, with the critical attention to rail corrugation in terms of the causes, measurements and solutions [9–11]. The rail corrugation with the fatigue crack in primary coil springs was tentatively correlated with structural vibration in experiments and simulations [12]. A linear model of the corrugation of rails was established to reveal the relationship between the initial wheel and rail roughness and the wear rate spectra in the contact patch [13]. The initiation and evolution of rail short-pitch corrugation and the effect on wheelset vibration were further investigated by using the field test and simulations to eliminate the corrugation [14].

Another argued aspect to buffer the fatigue crack is the wheel polygonalization traditionally found in high-speed railway wheels [15–18]. By considering the real wheel out-of-roundness, the considerably fluctuating wheel/rail contact force was observed from a vertical vehicle-track coupled dynamics model [17]. Similarly, by using on-site experiments and simulations, the mechanism of metro wheel polygonal wear was carefully explored at the speed of 65–75 km/h and suggested that the wheelset flexibility could accelerate the polygonal wear and abnormal vibration [19,20]. Such an irregular wheel profile was reasonably believed to produce a detrimental effect on the railway axle and bogie frame [21–23].

With regard to the crack of bogie frame [24,25], initial works have been made through multi-body dynamic system [26] and finite element simulations [27]. In this paper, on-track tests including the dynamic stress and acceleration were performed under different loading conditions to elucidate the cracking mechanism of metro bogie frames. By combining time, frequency and time-frequency domain analyses, the factors inducing the cracks are confirmed and measures are proposed to prevent such similar failures.

2. Experimental Methods

2.1. Cracking Site and Fracture Analysis

A crack was found at the spring sleeve of one metro bogie frame (as shown in Figure 1a) and the service life is less than 200,000 km. The spring sleeve of the frame side beam is welded by upper cover, lower cover and circular plate. The positioning block is welded to the spring sleeve lower cover by four welds, as seen in Figure 1b. The crack initiated from the weld toe of the positioning block and then propagated to the spring sleeve circular plate with maximum 110 mm in length, as shown in Figure 1c,d.

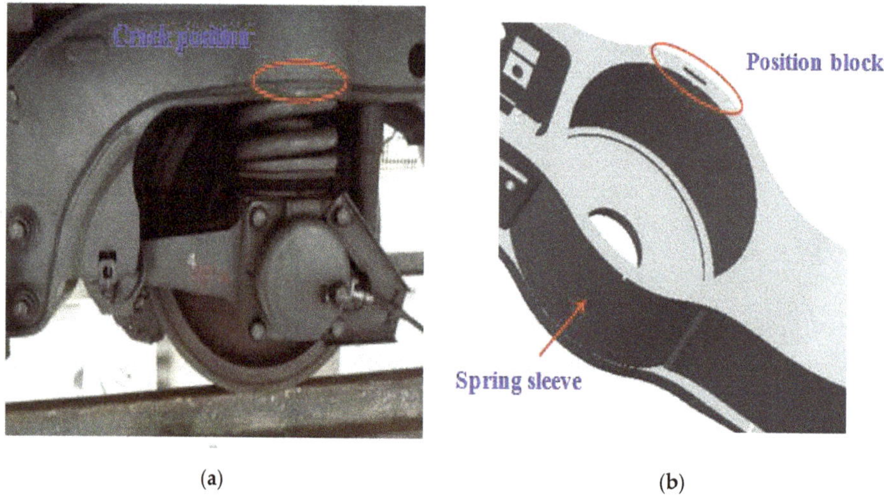

(a) (b)

Figure 1. *Cont.*

(c)   (d)

Figure 1. The crack position and the Macro morphology of the crack: (**a**) the crack position found in site; (**b**) the model of bogie frame and the positioning block; (**c**) and (**d**) the macro pictures of crack initiated from the weld toe of positioning block.

The fracture morphology of the cracking zone was examined by scanning electron microscopy (SEM) as shown in Figure 2. It can be clearly observed that some parallel marks are available near the weld toe, which indicates a notable fatigue damage feature during operation. Metallographic specimens were thus prepared at the critical cracking position. Columnar grains were clearly found in the weld and the heat-affected zone, which are composed of lamella ferrite and featherlike widmanstatten structures, as shown in Figure 3. The SEM observation of weld toe shows that the micro crack starts from the surface of the lower plate and propagates internally, and it includes several discontinuous sections and branches appear around the main crack (Figure 4).

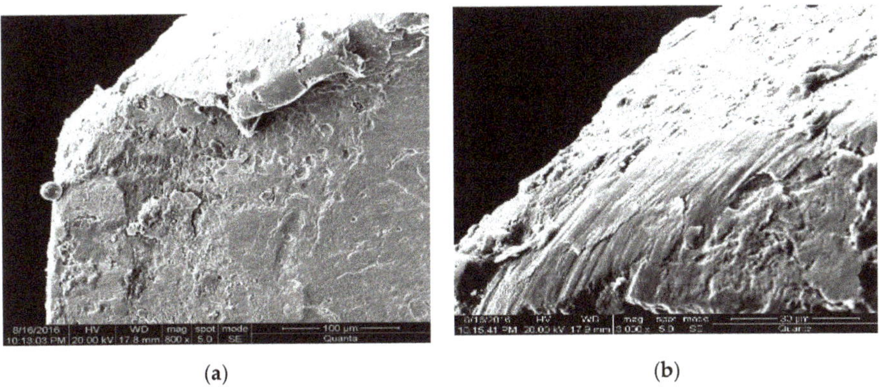

(a)   (b)

Figure 2. Micro-fracture morphology of the crack source and the fatigue striations: (**a**) 800×; (**b**) 3000×.

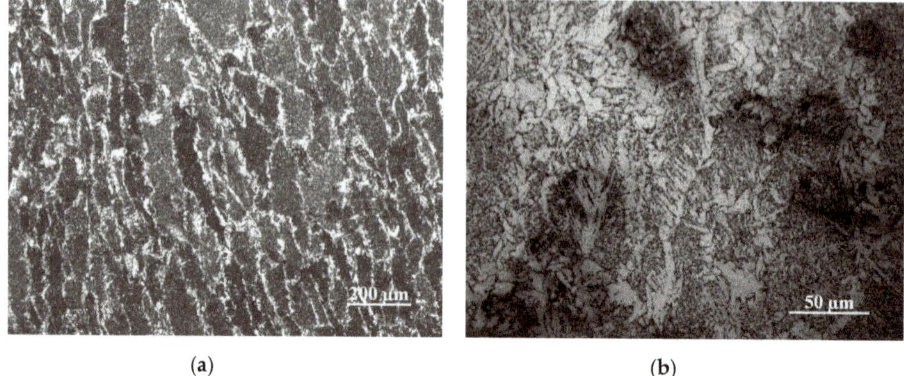

(a)   (b)

Figure 3. Microstructure of weld: (**a**) columnar grains enclosed by ferrite; (**b**) lamella ferrite and featherlike widmanstatten structure.

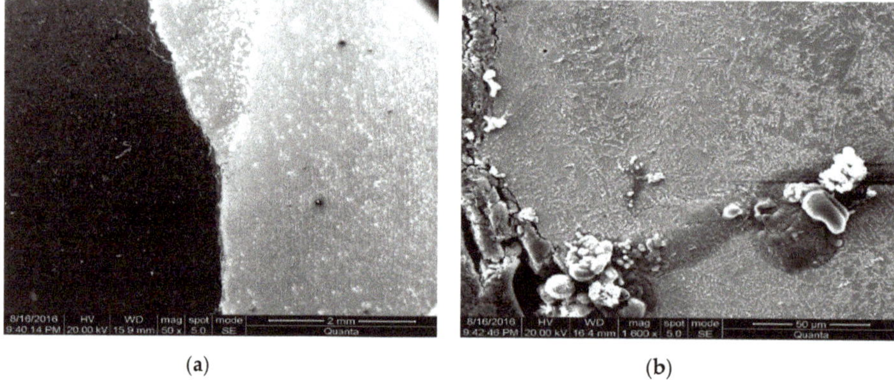

(a)   (b)

Figure 4. SEM of weld toe: (**a**) 50×; (**b**) 1600×.

2.2. On-Track Tests

In order to precisely acquire the failure causes and stress characteristics of the failed region in the metro vehicle bogie frame, both dynamic stress and vibration acceleration tests were firstly carried out under the following three operation conditions considering the wheel roundness, rail corrugation and the train running speed, as listed in Table 1.

Table 1. Testing conditions for dynamic stress and vibration acceleration.

No.	Test Conditions	Descriptions	Test Mileage/km
1	Normal operation	Bogie for normal operation	141.1
2	Normal operation	Bogie with profiled wheel	124
3	Interval speed limit operation	Bogie for normal operation	109

Digital dynamic signal acquisition system (IMC) was employed in this investigation due to its high precision and fast response during the whole process. The dynamic stress and acceleration sampling frequencies were set to 2500 Hz to ensure the completeness and reliability of the test data. The dynamic stress testing points (denoted such as D2, D11, D17) were mounted around the positioning block as a critical zone and on the lower cover of the side beam (such as D27), as shown in Figure 5. Note that points D02/ D11/ D17 were specially designed vertically to the weld toe of positioning block because mode I crack was generally assumed in engineering applications. The acceleration sensors were fixed on the axle box, spring sleeve and the motor, respectively, to measure the vibration induced by wheel-rail excitation, as shown in Figure 6.

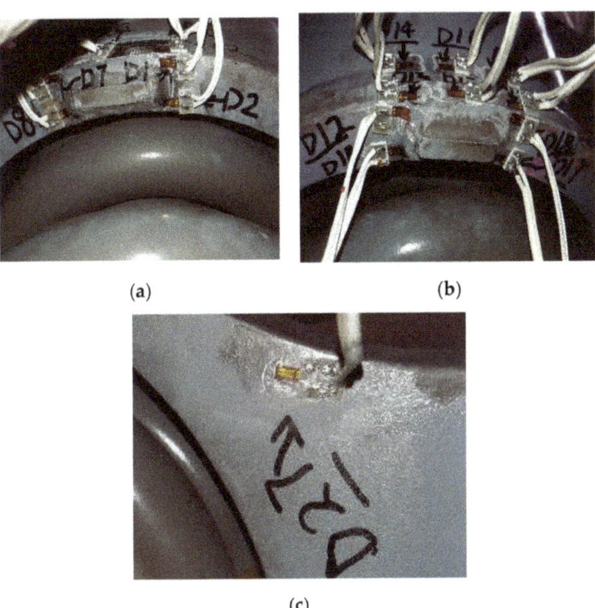

(a) (b)

(c)

Figure 5. Strain gauge tested points around the positioning block: (**a**) D02; (**b**) D11 and D17; (**c**) D27.

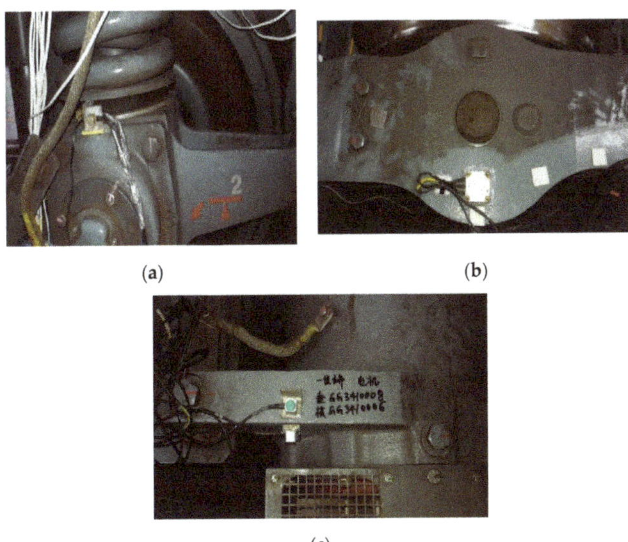

(a) (b)

(c)

Figure 6. Vibration acceleration tested points: (**a**) sensor on the axle box; (**b**) sensor on the spring sleeve; (**c**) sensor on the motor.

2.2.1. Dynamic Stress Test

The full-range dynamic stress waveforms of point D02, D11 and D17 are shown in Figure 7. The maximum stress amplitudes of D02, D11 and D17 are 91.5 MPa, 102.4 MPa and 85.6 MPa respectively. According to the stress-time history of each point, the stress spectrum can be obtained by adopting rain-flow counting method. Thus, with *S-N* curve and Miner's rule of accumulative damage, the damage of each point can be calculated. The results indicate that the damages of these testing points at the weld toe of positioning block are all larger than 1, which provide evidence of initiation and propagation of fatigue crack.

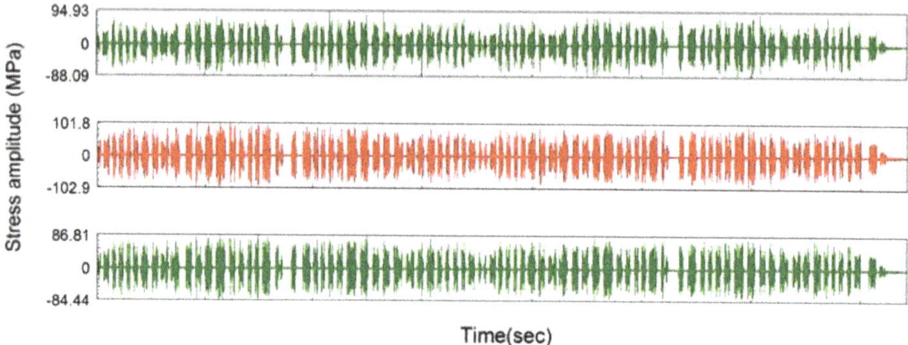

Figure 7. Full-range dynamic stress waveforms of D02, D11 and D17 (from top to bottom separately).

A typical stress time history signal of point D02 shown in Figure 8. The power spectral density (PSD) analysis and the speed-time-frequency relationship analysis of point D02 in some section are shown in Figures 9 and 10, respectively. It could be observed that the dominant frequencies are 71 Hz, 89 Hz and 94 Hz and strongly relevant to the operating speeds. This demonstrates that these frequencies are not the structural natural frequencies and there is no obvious behavior of the resonate vibration.

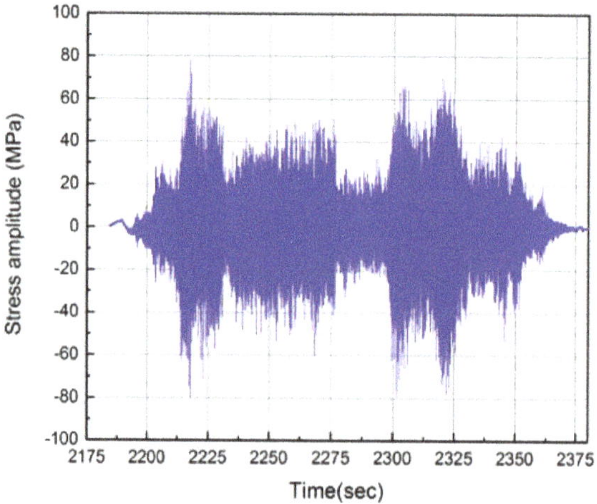

Figure 8. The stress waveform at point D02.

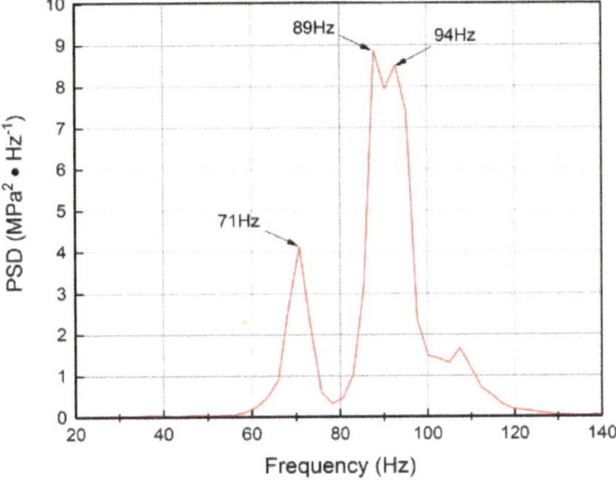

Figure 9. The power spectrum diagram at point D02.

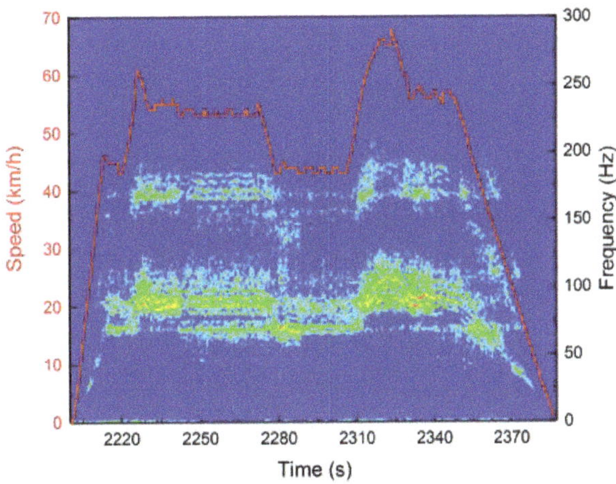

Figure 10. The speed-time-frequency relationship diagram at point D02.

2.2.2. Vibration Acceleration Test

The intrinsic relationship between the vibration acceleration and stress response of the bogie frame was conducted by analyzing the dynamic stress response at D02/D27 and the vertical accelerations of the axle box and frame, shown in Figure 11. It can be concluded that: (1) the vertical accelerations of the frame and the axle box are coincident with frequency bands at 71 Hz, 89 Hz and 94 Hz. Moreover, they have a significant following behavior in the time domain; (2) dynamic stresses of Point D02 and D27 are highly coherent with the vibration acceleration of the axle box and frame in the time domain and frequency domain, and Point D02 has much higher stress amplitude and PSD value due to the positioning block local stress concentration. The dynamic stress and acceleration analysis clearly indicate that the dynamic stress response at D02/D27 has obvious main frequencies and consistent with the speed. Combined with the vibration acceleration test results, the following conclusion can be drawn that the dynamic stress response at the positioning block weld toe is directly caused by the vibration of frame, which mainly derives from the wheel-rail excitation.

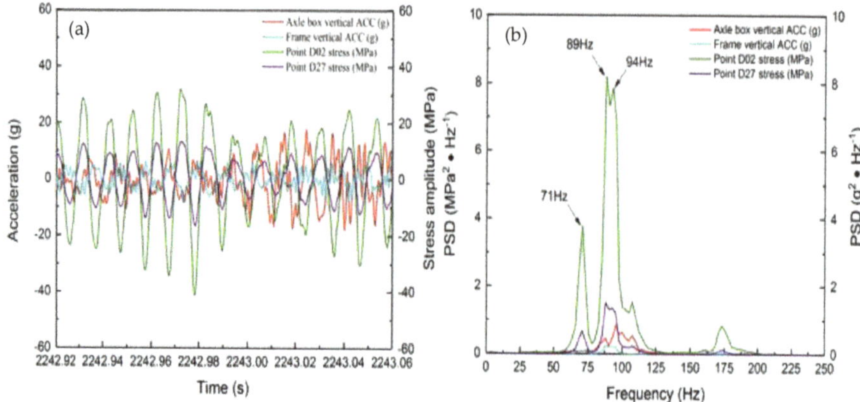

Figure 11. Relationship between the vibration acceleration and stress response: (**a**) time domain diagram; (**b**) power spectrum diagram.

3. Factors Relative to Crack Initiation and Propagation

3.1. Wheel Polygon

The roundness measurement results of No.1 and No.2 wheels installed on the tested bogie are shown in Figure 12a,b. No.1 wheel was a worn wheel after profiling, showing a good state. No.2 wheel was a worn wheel before profiling, demonstrating a significant polygonal worn phenomenon, with eccentric characteristics and 14- to 16-side polygon.

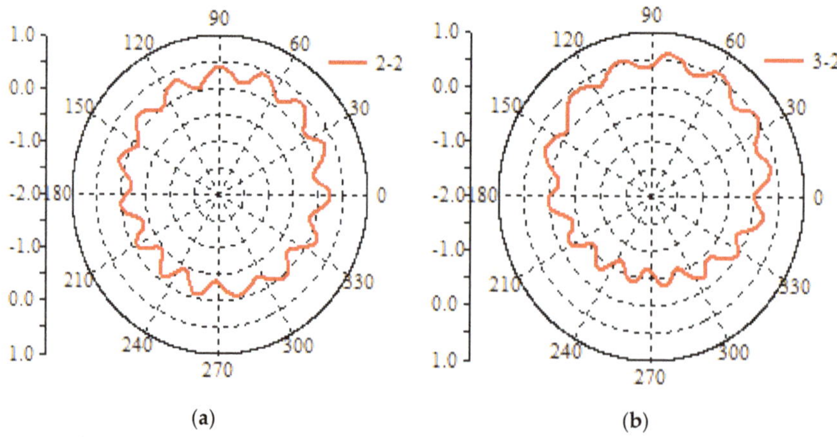

Figure 12. Polar diagram of roundness: (**a**) No.1 wheel; (**b**) No.2 wheel.

The wheel polygon will worsen the wheel-rail force and cause the high vibration of the axle box and bogie frame. The excitation frequencies of 14- to 16-side polygon wheel are coincided with the running speed linearly, shown in Figure 13. As indicated by the 15-sided wheel polygon analysis, for an example, when the operating speed is 54 km/h, the dominant frequency of the wheel-rail excitation is 86 Hz. Further, according to the analysis results of Figures 8 and 10, it can be observed that the dominant frequency of the wheel-rail excitation (86 Hz) is similar to the dominant frequency (89 Hz) of dynamic stress amplitude at point D02 in 54 km/h.

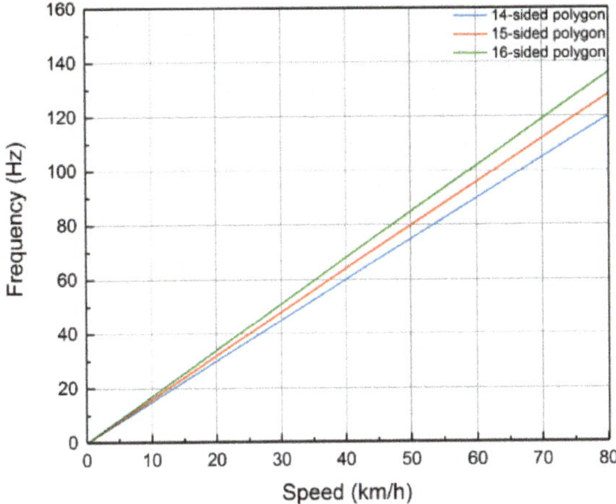

Figure 13. Relationship between wheel polygon and running speed.

Figure 14 indicates the comparison results between the time-frequency diagram at point D11 and the frequency excited by the wheel polygon. It can be observed that the characteristic frequency of 15-sides polygon (red points in Figure 13) is highly consistent with the dominant frequency of dynamic stress.

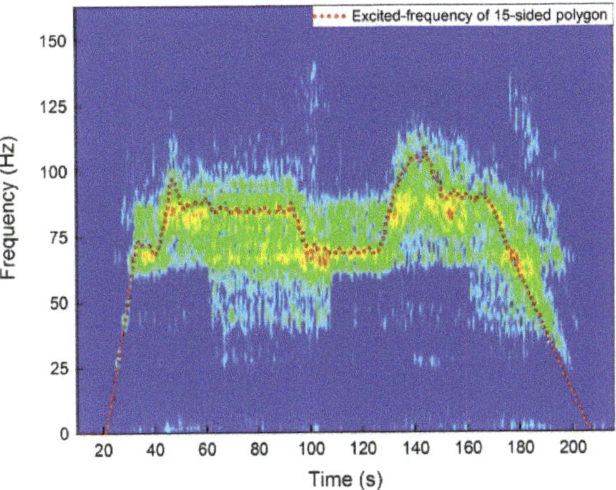

Figure 14. The frequency comparison between dynamic stress at point D11 and excitation from the wheel polygon phenomenon.

Figure 15 shows the comparison of the maximum dynamic stresses at D11 of the profiled and worn wheels. It can be observed that the maximum stress amplitude of the profiled wheel is significantly lower than that of the worn wheel, and the maximum reduction is 74%, indicating that the wheel polygon has a significant influence on the stress amplitude of the frame.

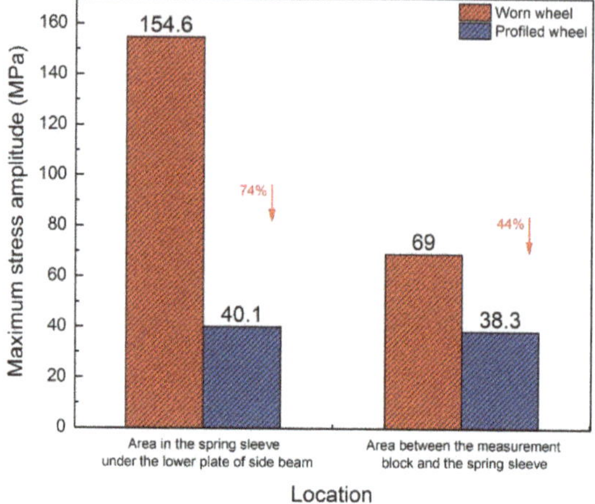

Figure 15. The comparison of the maximum stress amplitude of the profiled and worn wheels around the cracked area.

3.2. Rail Corrugation

The investigation of the rail condition indicates that the different rail types show different corrugation characteristics, and the obvious rail corrugation area is mainly occurred on the curve with the radius below 600 m (denoted as R600). Figure 16 exhibits the rail corrugation diagrams in different sections, which indicates significant rail corrugation phenomena on the rail surface.

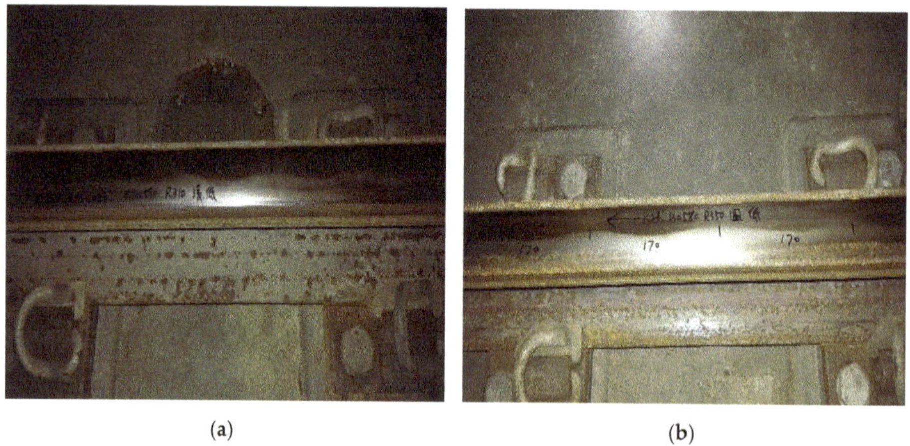

(a)   (b)

Figure 16. Rail corrugation in different sections: (a) section 1; (b) section 2.

Figure 17 illustrates the comparison of the dynamic stress in the cracked area with and without rail corrugation, where R320 section exists corrugation phenomenon but R400 section does not exist. The operating speeds are both 54 km/h. It can be observed that in the corrugation section, the maximum rail irregularity is 0.31 mm, and the maximum dynamic stress is 72.8 MPa. While in the non-corrugation section, the maximum rail irregularity is 0.18 mm, and the maximum dynamic stress is 41.6 MPa. Obviously, the dynamic stress amplitude of the measurement point with rail corrugation increases by 75% than that without rail corrugation.

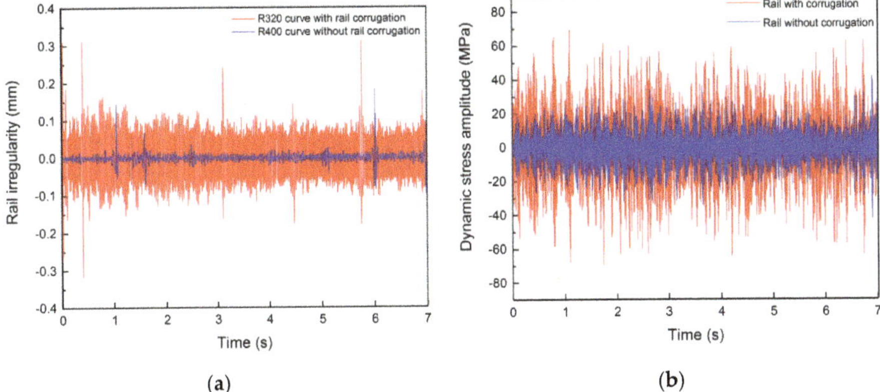

(a)   (b)

Figure 17. The relationship between rail irregularity and the dynamic stress amplitude: (**a**) degree of rail irregularity; (**b**) dynamic stress.

Meanwhile, Figure 18 shows a comparison of the characteristics of frequency domain with and without rail corrugation. It can be observed that in the frequency range of 60–90 Hz, the dynamic stress amplitude with rail corrugation is significantly greater than that without rail corrugation, especially in 71 Hz. In summary, the rail corrugation phenomenon has a significant effect on the dynamic stress amplitude at the cracked area.

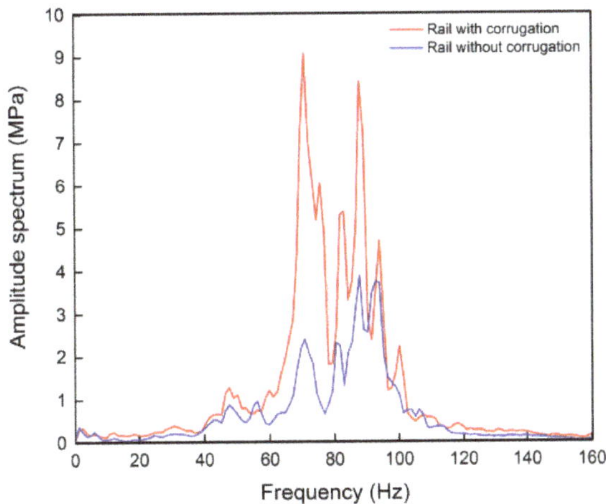

Figure 18. The amplitude spectrum comparison of the dynamic stress of the frame with and without the rail corrugation.

3.3. Running Speed

The wheel/rail vertical force increases with train running speed rising [17]. To investigate the relationship between the running speed and the dynamic stress amplitude, it is necessary to eliminate the influence of the rail condition. Therefore, a round-trip section was chosen to analyze. The round-trip speeds were 39 and 29 km/h, respectively, as shown in Figure 19. It can be observed that when the interval round-trip speeds were 29 and 39 km/h, respectively, the maximum values of dynamic stress amplitude were 10 and 36 MPa, and the main frequencies were 46 and 62 Hz. In conclusion, the amplitude and frequency of dynamic stress are significantly increased with the higher speed, which produces higher damage to the frame.

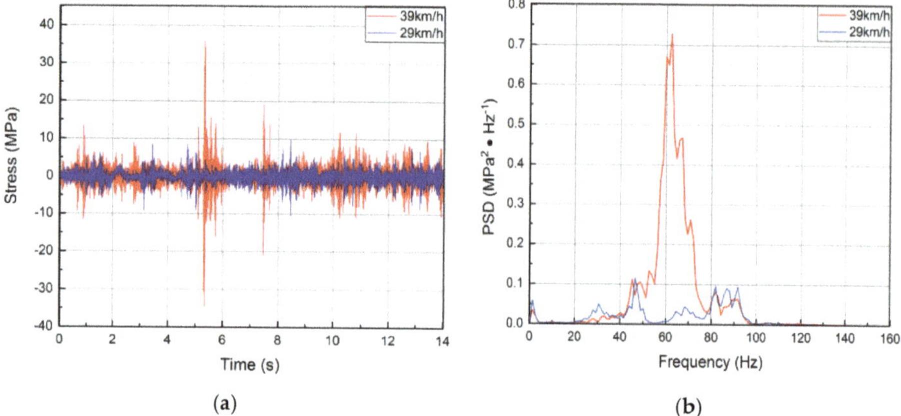

Figure 19. Round-trip time-frequency domain stress diagrams at different speeds: (**a**) time domain diagram; (**b**) time-frequency diagram.

4. Conclusions

An extensive investigation of the fatigue crack development of positioning block welded to the spring sleeve in metro bogie frame was conducted. Systematic studies of fracture and metallographic analysis were employed to obtain failure modes and fracture characteristics of the weld toe of positioning block. On-track testing was carried out to obtain acceleration and stress response information of the bogie. After that, the test data analysis in the time domain and the frequency domain was conducted. The results are as follows:

(1) The weld toe of positioning block was the cracking initiation site, in which micro cracks were observed.

(2) The vibration frequency generated by the wheel polygon and the rail corrugation is highly coherent with the dominant frequency of dynamic stress at the cracked area, indicating that the wheel polygon and the rail corrugation have a significant effect on the dynamic stress amplitude. Further, the response of the amplitude and frequency of dynamic stress is higher with running speed increasing.

(3) The fractography and the on-track tests show that optimizing the weld technology, improving the weld quality, profiling the wheels, grinding the rails and decreasing the train running speed are all effective to reduce the failure probability and elongate the bogie frame life.

Author Contributions: Conceptualization, W.J.W.; formal analysis, W.J.W.; methodology, W.J.W.; project administration, P.Y.Z.; resources, W.J.W. and J.Z.; writing original draft, J.Y.B.; writing—review and editing, W.J.W. and S.C.W. All authors have read and agreed to the published version of the manuscript.

Funding: This research was funded by Beijing Nova Program Interdisciplinary Cooperation Project, grant number Z191100001119011, National Natural Science Foundation of China, grant number 11572267 and Major Systematic Project of CHINA RAILWAY, grant number P2018J003.

Conflicts of Interest: The authors declare no conflict of interest.

References

1. Stanzl, S.E.; Tschegg, E.K.; Mayer, H.R. Lifetime measurements for random loading in the very high cycle fatigue range. *Int. J. Fatigue* **1986**, *8*, 195–200. [CrossRef]
2. Yang, B.; Duan, H.; Wu, S.; Kang, G. Damage tolerance assessment of a brake unit bracket for high-speed railway welded bogie frames. *Chin. J. Mech. Eng.* **2019**, *32*, 58. [CrossRef]

3. Guo, F.; Wu, S.C.; Liu, J.X.; Zhang, W.; Qin, Q.B.; Yao, Y. Fatigue life assessment of bogie frames in high-speed railway vehicles considering gear meshing. *Int. J. Fatigue* **2020**, *132*, 105353. [CrossRef]

4. Kassner, M. Fatigue strength analysis of a welded railway vehicle structure by different methods. International journal of fatigue. *Int. J. Fatigue* **2012**, *34*, 103–111. [CrossRef]

5. Wang, C.H.; Brown, M.W. Life prediction techniques for variable amplitude multiaxial fatigue—part 1: Theories. *Int. J. Fatigue* **1996**, *118*, 367–370. [CrossRef]

6. Lu, Y.; Xiang, P.; Dong, P.; Zhang, X.; Zeng, J. Analysis of the effects of vibration modes on fatigue damage in high-speed train bogie frames. *Eng. Fail. Anal.* **2018**, *89*, 222–241. [CrossRef]

7. Lučanin, V.J.; Simić, G.Ž.; Milković, D.D.; Ćuprić, N.L.; Golubović, S.D. Calculated and experimental analysis of cause of the appearance of cracks in the running bogie frame of diesel multiple units of Serbian railways. *Eng. Fail. Anal.* **2010**, *17*, 236–248.

8. Zhang, C.Y.; Cheng, Y.J. Random vibration fatigue analysis of ATP antenna beam based on measured spectrum data. *J. Dalian Jiaotong Univ.* **2015**, *36*, 44–47. (in Chinese).

9. Grassie, S.L. Rail corrugation: Characteristics, causes, and treatments. Proceedings of the Institution of Mechanical Engineers. *Part F J. Rail Rapid Transit* **2009**, *223*, 581–596. [CrossRef]

10. Sato, Y.; Matsumoto, A.; Knothe, K. Review on rail corrugation studies. *Wear* **2002**, *253*, 130–139. [CrossRef]

11. Zhang, H.; Liu, W.; Liu, W.; Wu, Z. Study on the cause and treatment of rail corrugation for Beijing metro. *Wear* **2014**, *317*, 120–128. [CrossRef]

12. Ling, L.; Li, W.; Foo, E.; Wu, L.; Wen, Z.; Jin, X. Investigation into the vibration of metro bogies induced by rail corrugation. *Chin. J. Mech. Eng.* **2017**, *30*, 93–102. [CrossRef]

13. Tassilly, E.; Vincent, N. A linear model for the corrugation of rails. *J. Sound Vib.* **1991**, *150*, 25–45. [CrossRef]

14. Yan, Z.Q.; Gu, A.J.; Liu, W.N.; Markine, V.L.; Liang, Q.H. Effects of wheelset vibration on initiation and evolution of rail short-pitch corrugation. *J. Cent. South Univ.* **2012**, *19*, 2681–2688. [CrossRef]

15. Meinke, P.; Meinke, S. Polygonalization of wheel treads caused by static and dynamic imbalances. *J. Sound Vib.* **1999**, *227*, 979–986. [CrossRef]

16. Brommundt, E. A simple mechanism for the polygonalization of railway wheels by wear. *Mech. Res. Commun.* **1997**, *24*, 435–442. [CrossRef]

17. Liu, X.; Zhai, W. Analysis of vertical dynamic wheel/rail interaction caused by polygonal wheels on high-speed trains. *Wear* **2014**, *314*, 282–290. [CrossRef]

18. Peng, B.; Iwnicki, S.; Shackleton, P.; Crosbee, D.; Zhao, Y. The influence of wheelset flexibility on polygonal wear of locomotive wheels. *Wear* **2019**, *432*, 102917. [CrossRef]

19. Cai, W.; Chi, M.; Tao, G.; Wu, X.; Wen, Z. Experimental and Numerical Investigation into Formation of Metro Wheel Polygonalization. *Shock Vib.* **2019**, *2019*, 1538273. [CrossRef]

20. Fu, B.; Bruni, S.; Luo, S. Study on wheel polygonization of a metro vehicle based on polygonal wear simulation. *Wear* **2019**, *438*, 203071. [CrossRef]

21. Wu, X.; Chi, M.; Gao, H. Damage tolerances of a railway axle in the presence of wheel polygonalizations. Engineering failure analysis. *Eng. Fail. Anal.* **2016**, *66*, 44–59. [CrossRef]

22. Wu, X.; Rakheja, S.; Qu, S.; Wu, P.; Zeng, J.; Ahmed, A.K.W. Dynamic responses of a high-speed railway car due to wheel polygonalisation. *Veh. Syst. Dyn.* **2018**, *56*, 1817–1837. [CrossRef]

23. Barke, D.W.; Chiu, W.K. A review of the effects of out-of-round wheels on track and vehicle components. *Proc. Inst. Mech. Eng. Part F J. Rail Rapid Transit* **2005**, *219*, 151–175. [CrossRef]

24. Fu, D.; Wang, W.; Dong, L. Analysis on the fatigue cracks in the bogie frame. *Eng. Fail. Anal.* **2015**, *58*, 307–319. [CrossRef]

25. Qin, G.D.; Liu, Z.M. Practical analysis method for fatigue life of welded frame of speed-up bogie. *China Railw. Sci* **2004**, *25*, 46.

26. Dietz, S.; Netter, H.; Sachau, D. Fatigue life prediction of a railway bogie under dynamic loads through simulation. *Veh. Syst. Dyn.* **1998**, *29*, 385–402. [CrossRef]

27. Kim, J.S. Fatigue assessment of tilting bogie frame for Korean tilting train: Analysis and static tests. *Eng. Fail. Anal.* **2006**, *13*, 1326–1337. [CrossRef]

© 2020 by the authors. Licensee MDPI, Basel, Switzerland. This article is an open access article distributed under the terms and conditions of the Creative Commons Attribution (CC BY) license (http://creativecommons.org/licenses/by/4.0/).

Article

Thermal Fatigue Life Prediction of Thermal Barrier Coat on Nozzle Guide Vane via Master–Slave Model

Peng Guan [1], Yanting Ai [2], Chengwei Fei [3,*] and Yudong Yao [2]

[1]   School of Power and Energy, Northwestern Ploytechnical University, Xi'an 710072, China; 613511gp@163.com
[2]   Faculty of Aviation Engine, Shenyang Aerospace University, Shenyang 110136, China; ytai@163.com (Y.A.); yaoyudong148@163.com (Y.Y.)
[3]   Department of Aeronautics and Astronautics, Fudan University, Shanghai 200433, China
*   Correspondence: cwfei@fudan.edu.cn

Received: 16 September 2019; Accepted: 11 October 2019; Published: 16 October 2019

Abstract: The aim of this paper was to develop a master–slave model with fluid-thermo-structure (FTS) interaction for the thermal fatigue life prediction of a thermal barrier coat (TBC) in a nozzle guide vane (NGV). The master–slave model integrates the phenomenological life model, multilinear kinematic hardening model, fully coupling thermal-elastic element model, and volume element intersection mapping algorithm to improve the prediction precision and efficiency of thermal fatigue life. The simulation results based on the developed model were validated by temperature-sensitive paint (TSP) technology. It was demonstrated that the predicted temperature well catered for the TSP tests with a maximum error of less than 6%, and the maximum thermal life of TBC was 1558 cycles around the trailing edge, which is consistent with the spallation life cycle of the ceramic top coat at 1323 K. With the increase of pre-oxidation time, the life of TBC declined from 1892 cycles to 895 cycles for the leading edge, and 1558 cycles to 536 cycles for the trailing edge. The predicted life of the key points at the leading edge was longer by 17.7–40.1% than the trailing edge. The developed master–slave model was validated to be feasible and accurate in the thermal fatigue life prediction of TBC on NGV. The efforts of this study provide a framework for the thermal fatigue life prediction of NGV with TBC.

Keywords: thermal fatigue; thermal barrier coat; master–slave model; life prediction; nozzle guide vane

1. Introduction

With the improvement of aero-engine performance with a high flow rate and high thrust–weight ratio, the temperature and pressure of gas at the outlet of the combustion chamber is rising. For instance, the temperature before the turbine for the new generation engines (e.g., F119, F120, EJ200, etc.) is over 1850 K. To meet the harsh working environment, a thermal barrier coat (TBC) is a key technology in protecting turbine blades (rotor blades and nozzle guide vanes (NGVs)) [1–3]. As the first stage NGV faces high combustion outlet temperatures, the TBC at the surface of the NGVs endures a high thermal cycle load, so its failure problems are particularly serious due to the thermal loads. Therefore, it is necessary to consider the thermal fatigue life (TFL) of the TBC to improve the design and machining of the first-stage NGV [4].

TBC life is seriously induced by four components as shown in Figure 1 [5]: (i) Top coat (TC) layer: providing thermal insulation; (ii) Thermally grown oxide (TGO) layer: providing the bonding of TBC with bond coat (BC) to slow subsequent oxidation; (iii) BC layer: containing the source of elements to create TGO in oxidizing environment and provide oxidation protection; and (iv) superalloy substrate: carrying mechanical loads. Through the oxidation of high-temperature gas, the aluminum in the BC is oxidized to produce an alumina layer (also called the TGO). With the increase in the TGO thickness, the material properties between the TC and BC are destroyed. The mixed layer TC/TGO is liable to crack and peel off under the combined effect of thermal stress and deformation in different layers, which eventually leads to the failure of the NGV [6]. Therefore, it is of some urgency to study the mechanical properties of a mixed layer TC/TGO with regard to a thermal shock environment. Robert [7] presented a Thermo-Calc/Dictra-based approach for the life prediction of isothermally oxidized atmospheric plasma sprayed TBCs. The beta-phase depletion of the coating was predicted and compared to the life prediction criteria based on the TGO thickness and aluminum content in the coating. Based on the BC test results, the life of the TBC was predicted by Song [8], according to degradation and thermal fatigue. Many experiments indicated that the failure of the TBC systems under thermomechanical loading was complicated due to the influences of many factors such as thermal mismatch, oxidation, interface roughness, creep, sintering, and so forth [9–11].

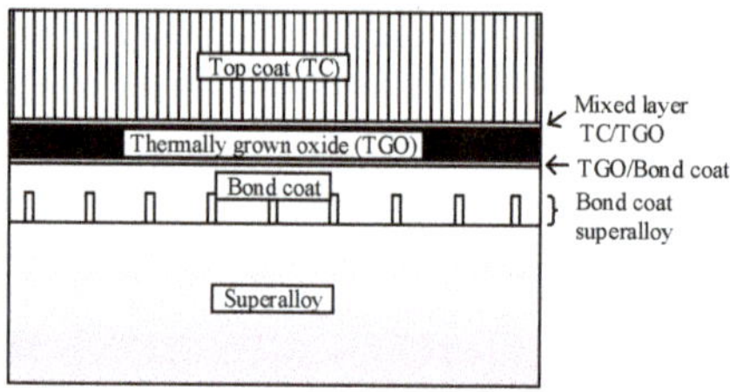

Figure 1. Thermal barrier coating and its four components. TC: Top coat; TGO: Thermally grown oxide.

Among the failure factors, the effect of TGO growth is particularly prominent, because volume expansion and compressive stress increase are correlative with the growth of TGO. When the NGV gets to the cooling state, the induced unbalanced temperature at the mixed layer TC/TGO leads to a high residual compressive stress, which varies directly with the strain energy. When the strain energy reaches the limit value, cracks occur. Yang et al. [12,13] tested an oxidation weight gain of the TBC at 1323 K and obtained the dynamic test curves for characterizing the TGO thickness. Based on this work, Wei et al. [14] established a TFL model of a TBC regarding the phenomenological approach and low cycle fatigue theory. However, the working environment and micro-strain characteristics of the NGV were not discussed in detail. In the process of the thermal shock cycle, the flow, heat transfer, and structure interact with each other, thus it is difficult to separate the temperature field from thermal stress in the calculation of flow characteristics. Therefore, the TFL of a NGV is a typical fluid-thermo-structure (FTS) coupling problem.

With the development of numerical simulation approaches, it is possible to calculate the thermal shock cyclic loads of a NGV by combining computational fluid dynamics (CFD) and finite element (FE) methods. Kim et al. [15] studied the heat transfer coefficients and stresses on blade surfaces using the finite volume (FV) and FE methods and obtained the maximum material temperature and thermal stress at the trailing edge near mid-span. Meanwhile, he also discussed the life prediction methods of turbine components by coupling aero-thermal simulation with a nonlinear thermal-structural FE model and a slip-based constitutive model [16]. Chung et al. [17] predicted the cracks on the vane of a power generation gas turbine by the FTS method and Guan et al. [18] carried out a simulation investigation of temperature variation, and obtained the stress and vibration characteristics of a NGV by the FTS method.

Although the FTS technology can reveal the physical characteristics of a NGV in the thermal cycle under macroscopic scales, the thermal fatigue problems of TBC gradually emerge from micro-cracks [19]. Up to now, it is possible to establish a mathematical grid model with respect to both the macro and micro scales. However, the solution of the grid model requires huge computing resources and time consumption, which seriously restricts the development of new thermal barrier coated NGVs [20]. Therefore, to provide an effective TFL model of the TBC, it is necessary to integrate flow, heat transfer, and structure with the micro- and macro-scale to perform an integrity design. The concept of high-integrity was proposed by the United States Air Force in 2012 for the numerical simulation of a NGV with a TBC [21]. High-integrity requires the consistency of the numerical simulation with a real structure. Due to the costly computation, however, we have not found related works to highly-integrated TBC simulation so far.

Along with the heuristic thought, it is unsatisfactory for a highly-integrated analysis of the life prediction of a NGV with a TBC to only consider heat shock performance in the macro- or micro-scale. This paper attempts to develop a high-integrity approach, in other words, the master–slave model involving the FTS coupling technology, phenomenological life prediction method, and volume mesh mapping algorithm for TBC thermal fatigue life prediction and the improvement in prediction precision and efficiency. Here, the master–slave model was used to establish the relationship between the macro-scale and micro-scale in the thermal fatigue life prediction of a TBC.

The remainder of this paper is organized as follows. Section 2 introduces the modeling and simulation approaches including the physical model, material parameters, boundary conditions, meshing, and simulation procedure. The verification strategy and establishment of thermal fatigue life are discussed in Section 3. In Section 4, the results and discussion involving the heat transfer analysis, temperature filed analysis, thermo-structural analysis, and thermal fatigue life prediction of the TBC on a NGV are investigated. Section 5 gives the conclusions and findings of this paper.

2. Modeling and Simulation Methods

2.1. Thermal Fatigue Life Theory of Thermal Barrier Coat

To establish the TFL prediction model of the TBC, the classical phenomenological life prediction method proposed by Meier et al. [22] was referred to in this paper. According to the theory, the TFL of the TBC is related to the TGO thickness and strain range in the thermal cycle. The experimental correlation of the TGO layer thickness is described as:

$$\delta = \left\{ \exp\left[Q\left(\frac{1}{T_0} - \frac{1}{T} \right) \right] \cdot t \right\}^n \tag{1}$$

where δ, t, and T indicate the TGO thickness, oxidation time, and oxidation temperature, respectively; Q, T_0, and n are the active energy of oxygen atom diffusion in metals, undetermined temperature, and undetermined coefficient, which are 28,230, 1572, and 0.313 at 1320 K, respectively, as determined by the experimental investigation of a plasma sprayed TBC material [13].

The phenomenological TFL model is

$$N = \left[\left(\frac{\Delta\varepsilon_{f_0}}{\Delta\gamma_i} + a\Delta\varepsilon_i \right)\left(1 - \frac{\delta}{\delta_c} \right) + \left(\frac{\delta}{\delta_c} \right) \right]^b \tag{2}$$

where N indicates the number of cycles; $\Delta\gamma_i$ and $\Delta\varepsilon_i$ are the shear strain range and axial strain range at risk point in the ith thermal cycle, respectively; $\Delta\varepsilon_{f_0}$ is the critical tensile strain range; δ_c is the critical TGO thickness; and a and b are unknown parameters in Equation (2).

Regarding $\Delta\varepsilon_{f_0} = 0.087$ and $\delta_c = 0.058$ mm for the plasma sprayed TBC material [14], Equation (2) is considered as a time-dependent oxidation-induced cracking process accompanied by oxidation failure at the mixed TC/TGO layer with respect to the peel off process of the TBC. Miner's linear cumulative damage model [22] was used to compute cyclic life in this paper.

In cumulative theory, the damage caused by one cycle is assumed to be $D_m = 1/N_m$. The total damage D caused by multiple cycles is

$$D = \sum_{m=1}^{k} D_m = \sum_{m=1}^{k} \frac{1}{N_m} \tag{3}$$

where m denotes the number of cycles; N_m is the life under m cycles; and k indicates the maximum number of cycles. When the damage D is larger than 1, the TBC is considered to have failed.

2.2. Simulation Procedure

The high-integrity life prediction approach for the TBC on a NGV with the master–slave model is shown in Figure 2.

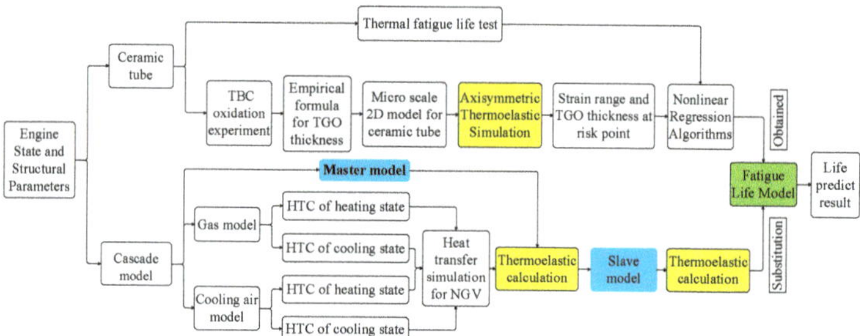

Figure 2. High-integrity life predication approach for the thermal barrier coat (TBC) on NGV.

To find the coefficients (a and b) in Equation (2), we need the TFL test data of the NGV. Due to the insufficient TFL measurement data of the NGV in an open database [23], the TFL of a typical ceramic metal tube, a similar plasma sprayed TBC material, and the working environment of the model described in this paper, was selected to calculate the value of parameters a and b in Equation (2). The ceramic tube test has been proven to be an effective alternative test method [13]. The prediction procedure of the TFL for the TBC is described as follows:

Step 1: Based on the geometry parameters and test conditions of a ceramic metal tube, a 2D axisymmetric FE model was established and solved by the thermo-elastic method. Both shear strain ranges and axial strain ranges (i.e., strain ranges in the Z direction) at the risk point in one cycle were simulated with different TGO thicknesses, and the relationship of the strain and thickness was obtained by fitting polynomials. The relationship was inputted into Equation (2). According to the cyclic life test of the ceramic metal tube, the parameters a and b were obtained by nonlinear regression analysis.

Step 2: The typical thermal shock test of the NGV was divided into two states (i.e., heating stage and cooling stage). The convection heat transfer coefficient of each stage was solved by the CFD method, and then imported into the master model (i.e., macro NGV with TBC) as the boundary conditions. Then, the temperature and node deformations were solved by the thermo-elastic coupling method. The results at each time were inputted into the slave model (i.e., micro TBC) by the volume element intersection mapping algorithm [24]. The master–slave model was based on two assumptions: (a) The TGO hardly affects the temperature field in the TBC, and (b) the grid node deformation mainly depends on the temperature field at the macro scale [25].

Step 3: The strain ranges at the risk points of the slave model were imported into the TFL model to predict the cyclic life of TBC.

2.3. Physical Model and Meshing Method

The focus of this paper was on the first stage NGV of an axial flow turbojet engine [26]. The flow cascade was comprised of 24 vanes. The internal and external radii of the link rings were 95 mm and 135 mm, respectively. In this study, the fan-shaped cascade was simplified to the square cascade with a vane space of 38 mm, while both cooling holes and link rings were ignored. The geometrical size of the NGV is shown in Figure 3. As illustrated in Figure 3, the geometry of the master model consisted of three layers (i.e., TC (material mode no. 8YSZ with thickness 0.25 mm), BC (material mode no. NiCrAlY with thickness 0.125 mm), and Ni-based alloy substrate (material mode no. GH3030)).

The TGO layer only existed in the slave model to calculate the strain range for different thicknesses. The master–slave model is shown in Figure 4a.

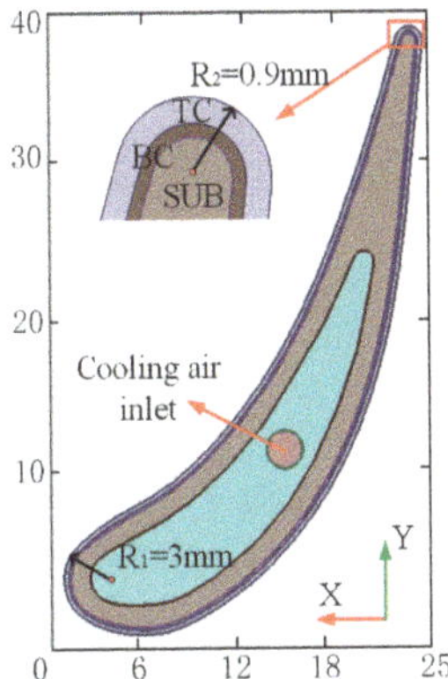

Figure 3. Section view (mm).

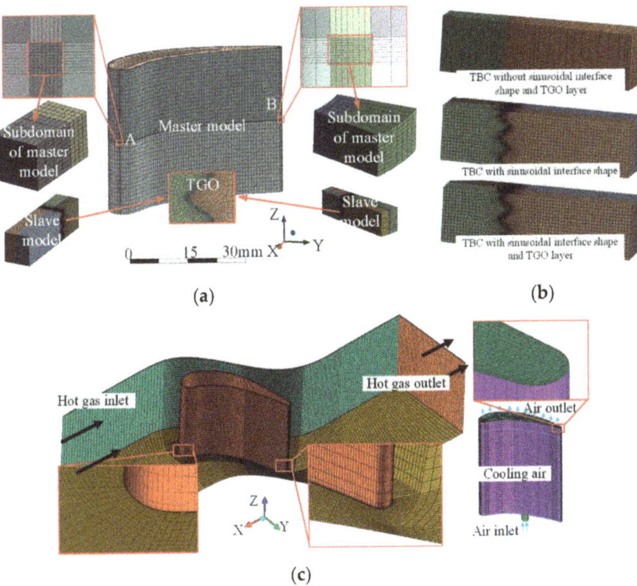

Figure 4. Meshing of the NGV and master–slave models. (**a**) Master–slave model; (**b**) Typical slave models; (**c**) Flow domain.

Based on the morphological characteristics of the TGO layer [27], a sinusoidal interface, as indicated in Figure 4b, was adopted to approximate the TGO and mixed layer with the wavelength of 0.04 mm and amplitude of 0.01 mm. Three kinds of slave models were established, as seen in Figure 4b, for the TBC with a flat surface, TBC with a sinusoidal interface, and TBC with a TGO layer. As high-temperature regions are usually at the center of the leading edge and trailing edge, two pieces of the TBC layer were defined as the subdomain of the master model in the positions. To prevent a large volume ratio of grid elements in the master model, the subdomain was slightly larger than the slave model.

According to the structure of the master model, the gas and cooling air models were meshed. The thickness of the first layer near the wall grids was 1 μm and the expansion ratio was 1:2. The grids of gas, cooling air, and master model were selected by mesh independence tests with respect to the sizes of the cell of 520,000, 40,000, and 330,000, respectively, and the grid numbers of the subdomain and slave model were 10,000 and 580,000, respectively. All grid models are shown in Figure 4. A fully coupling element, ANSYS-Solid 226 (Provided by workbench 18.0 software, ANSYS, Pittsburgh, PA, USA), was adopted in the master model. The fully coupling technology can improve the accuracy of thermo-elastic coupling by 2% to 3% with regard to temperature displacement coupling theory [28]. As the temperature was already solved in the master model, a weak coupling element, ANSYS-Solid 186, was adopted in the slave model. As for the master–slave model, it is unnecessary to repeatedly build and simulate the FE model with the increase in the TGO thickness. Thus, it is necessary to improve the work efficiency of the TFL prediction.

2.4. Material Parameters and Boundary Conditions

In the above FE model, each layer of the NGV and TBC is considered as an isotropic and homogeneous material. The TC is modeled as an elastic body, whereas the BC, TGO, and substrate are regarded as elastic–perfectly plastic materials [29,30]. The temperature-dependent material properties are listed in Tables 1 and 2. It should be noted that to limit stress to the experimental level [31,32], the TGO is allowed to undergo stress relaxation at high temperatures, which is realized by introducing the yield strength of TGO at peak temperatures [31]. The multilinear kinematic hardening model was selected to simulate the elastic–plastic behaviors. The back-stress tensor for multilinear kinematic hardening evolves such that the effective stress versus effective strain curve is multilinear with linear segments defined by the stress–strain–temperature points [33]. The yield stress σ_{yi} ($i = 1, 2, \ldots, 7$) for the *i*th temperature point in Table 2 is

$$\sigma_{yi} = \frac{1}{2(1+v)}(3E\varepsilon_i - (1-2v)\sigma_i) \tag{4}$$

where (ε_i, σ_i) is stress and strain at the *i*th temperature point, respectively; E is the Young modulus; and v is the Poisson ratio.

Table 1. Temperature dependent material parameters for different layers [30].

Parameters	SUB	BC	TGO	TC
Temperature, K	20–1100	20–1100	20–1100	20–1100
Thermal expansion coefficient, 10^{-5}/K	1.48–1.80	1.36–1.76	0.80–0.96	0.90–1.22
Young modulus E, GPa	220–120	200–110	400–320	48–22
Poisson ratio v	0.31–0.35	0.30–0.33	0.23–0.25	0.10–0.12
Shear modulus, $\times 10^{11}$ Pa	0.84–0.44	0.77–0.41	1.66–1.28	0.21–0.01
Heat transfer coefficient, W/m·K	88–69	5.8–17	10–4.1	2–1.7
Density, kg/m^3	8500	7380	3984	3610
Specific heat, J/kg K	440	450	755	505

Table 2. The variations of yield strength with temperature for different layers [31].

No.	Temperature	SUB	BC	TGO
1	300 K	800 MPa	426 MPa	10,000 MPa
2	473 K	800 MPa	412 MPa	10,000 MPa
3	673 K	800 MPa	396 MPa	10,000 MPa
4	873 K	800 MPa	362 MPa	10,000 MPa
5	1073 K	800 MPa	284 MPa	10,000 MPa
6	1273 K	800 MPa	202 MPa	1000 MPa
7	1373 K	800 MPa	114 MPa	1000 MPa

Table 3 reveals the performance parameters of the axial flow turbojet engine, which are employed to derive the simulated boundary conditions of the master model.

Table 3. Engine performance parameters [26].

Parameters	Value
Inlet average total temperature T^{*}_{eng}, K	290.29 K
Exhaust nozzle total temperature T^{*}_{ex}, K	917.33 K
Inlet static pressure P_{eng}, KPa	95.22 KPa
Compressor outlet total pressure P^{*}_{com}, KPa	538.39 KPa
Inlet air flow of the engine W_{eng}, g/s	3520 g/s
Fuel mass flow W_{fu}, g/s	54.48 g/s
Mach number of engine inlet Ma_{eng}	0.31

In line with the power balance principle of the aeroengine rotor shaft, the relationship between the inlet Mach number Ma_{eng} and the engine inlet total pressure P^{*}_{eng} is

$$P^{*}_{eng} = P_{eng}\left(1 + 0.2 \cdot Ma_{eng}^{2}\right)^{3.5} \tag{5}$$

where P_{eng} is the inlet static pressure.

With respect to the efficiency of compressor $\eta_{com} = 0.89$ [34], the outlet temperature of compressor T^{*}_{com} is

$$T^{*}_{com} = T^{*}_{eng}\left(1 + \frac{\pi^{0.2857} - 1}{\eta_{com}}\right) \tag{6}$$

where $\pi = P^{*}_{com}/P^{*}_{eng}$ is the pressure ratio of the compressor and T^{*}_{eng} is the total temperature of engine inlet.

The temperature of the cooling air is determined by the outlet temperature of the compressor because cooling air is sucked up directly from the end of compressor. In light of Equations (4)–(6), the cooling air mass flow W_{c}, mean temperature T_{mean} at the cascade inlet, and inlet total pressure P_{in}^{*} are computed by

$$W_{c} = W_{eng}\left[(1 - w)(1 + F) + w\right] \tag{7}$$

$$T_{mean} = \frac{C_{g}W_{eng}T^{*}_{com}}{C_{pg}W_{c}\eta_{sh}} + T^{*}_{ex} \tag{8}$$

$$P^{*}_{in} = P^{*}_{com} \cdot \eta_{rs} \tag{9}$$

where $w = 1.88\%$ is the percentage of coolant flow; $F = W_{fu}/W_{eng}$ is the gas–oil ratio, where W_{fu} is the oil mass and W_{eng} is the gas mass flow; C_{g} is the specific heat capacity of air; C_{pg} is the specific heat capacity of hot gas; $\eta_{sh} = 99\%$ is the mechanical efficiency of aeroengine shaft; $\eta_{rs} = 0.97$ is the total pressure recovery coefficient of combustor [34]; T^{*}_{ex} indicates exhaust temperature; and P^{*}_{com} expresses the inlet total pressure of compressor.

In terms of Equations (5)–(9), the conditions of the NGV simulation are shown in Table 4.

Table 4. Simulation conditions of the NGV.

Variable	Value
Inlet total pressure, P_{in}^{*}	522 KPa
Inlet average temperature, T_{mean}	1310 K
Outlet mass flow, W_{out}	148.9 g/s
Cooling air mass flow, W_{ca}	2.8 g/s
Cooling air inlet total temperature, T^{*}_{com}	491 K

All boundaries were set as no-slip walls, and the periodic boundary method was used to predict periodic flow. The influence of gas kinetic energy on heat transfer is considered by the total energy model. The selected turbulence model SST γ-θ, which is usually used in the simulation of cascades, has a high prediction accuracy [35]. The solution scheme of the model is the second-order backward

Euler method [36]. When the residue error of mass flow is less than 10^{-6}, the calculation is considered to be convergent. The bottom surface of the NGV was restrained in the z direction and the top surface was constrained in the x and y directions.

3. Life Modeling Process

The TFL test data of the ceramic metal tube were adopted to build the phenomenological TFL model (Equation (2)) by finding parameters a and b. The inner diameter, outer diameter, and length of the ceramic metal tube were 11 mm, 15 mm, and 85 mm, respectively. In the TFL test experiment, the central region of the tube with the length of 50 mm was heated for 160 s by an electromagnetic induction coil. After heating, the inner surface was cooled for 260 s by high pressure air [13]. The measured interface temperature of the TBC in a single cycle are shown in Figure 5. The TFL of a typical NGV was tested under six operating conditions (case 1, case 2, … , case 6). The results of the tests are shown in Table 5.

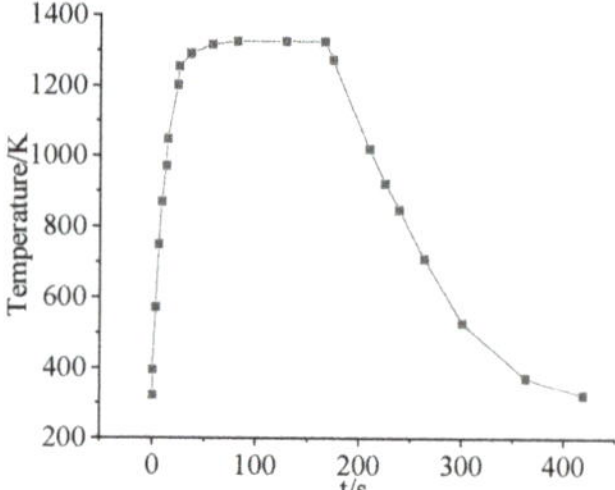

Figure 5. Interface temperature of the TBC in a single cycle [13].

Table 5. Test results of the thermal fatigue experiment under a preheating temperature of 1323 K [14].

Number	Test Condition		Cycle
	Preheating Time, H	Heat Preservation Time, S	
Case 1	0	0	298
Case 2	0	670 s	505
Case 3	50 h	0	480
Case 4	100 h	0	441
Case 5	200 h	0	400
Case 6	300 h	0	206

The 2D axisymmetric FE model was established by adopting a micro-segment with the length of 0.125 mm in the center of the heating part, as shown in Figure 6. In the 2D FE model, the size, structure, and material of the TBC layers were consistent with the slave model, and the thermal cyclic load and heating/cooling period were also the same as those in the test. We separated the measured temperature to the cooling temperature curve and heating temperature curve, which were assigned to the inner boundary and outer boundary of the 2D axisymmetric finite element model, respectively. Displacement constraints in the z direction and x direction were applied to the lower edge and inner edge, respectively. A fully coupling element, ANSYS-Plane 223, was used to simulate the transient temperature and thermal stress based on the thermoelastic damping matrix, i.e.,

$$\begin{bmatrix} [M] & [0] \\ [0] & [0] \end{bmatrix} \begin{Bmatrix} \{\ddot{u}\} \\ \{\ddot{T}\} \end{Bmatrix} + \begin{bmatrix} [C] & [0] \\ [C^{tu}] & [C^{t}] \end{bmatrix} \begin{Bmatrix} \{\dot{u}\} \\ \{\dot{T}\} \end{Bmatrix} + \begin{bmatrix} [K] & [K^{ut}] \\ [0] & [K^{t}] \end{bmatrix} \begin{Bmatrix} \{u\} \\ \{T\} \end{Bmatrix} = \begin{Bmatrix} \{F\} \\ \{Q\} \end{Bmatrix} \tag{10}$$

where $[M]$, $[C]$, and $[K]$ are the matric of element mass, element structural damping, and element stiffness, respectively; $\{u\}$ and $\{T\}$ are the vector of displacement and temperature, respectively; $\{F\}$ indicates the um of the element nodal force; $[C^{t}]$ and $[K^{t}]$ are the matric of element specific heat, element thermal conductivity, respectively; $\{Q\}$, $[K^{ut}]$, and $[C^{tu}]$ denote the sum of the element heat generation load and element convection surface heat flow, the matrix of element thermoelastic stiffness, and the matrix of element thermoelastic damping, respectively.

Six typical TGO thicknesses (0, 2 μm, 4 μm, 6 μm, 8 μm, and 10 μm) were selected in the simulation. To find the risk time and point in the thermal cycle, the von-Mises stress was selected, i.e.,

$$\sigma_{\text{von}} = \left[\frac{\left(\sigma_x - \sigma_y\right)^2 + \left(\sigma_x - \sigma_z\right)^2 + \left(\sigma_y - \sigma_z\right)^2}{2}\right]^{\frac{1}{2}} \tag{11}$$

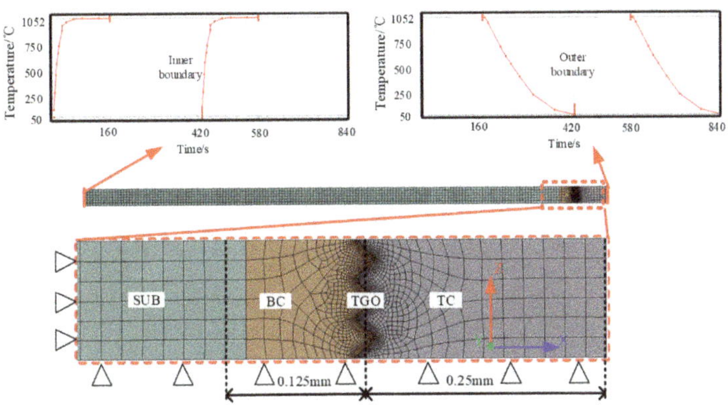

Figure 6. Finite element (FE) model and boundary conditions of the ceramic metal tube.

The maximum thermal stress of the TC layer is shown in Figure 7a. As seen in Figure 7a, the change in thermal stress basically agrees with the temperature curve in Figure 5, and the thermal stress increased with the increasing TGO thickness. The maximum thermal stress occurred in 50 s to 160 s in the first cycle. For the strain contour at the thickness of 2 μm, the shear strain and axial strain are shown in Figure 7b. The results in Figure 7b show that the maximum shear strain and axial strain are near by the bottom of the valley (i.e., point I and II, respectively). The cracks caused by shear strain were perpendicular to the mixed layer of the TGO, while the cracks induced by axial strain were tangent to the valley center (z direction in Figure 6). The value of the shear strain was higher than that of the axial strain, thus the grid node of shear strain was regarded as the risk point for the simulation. The axial and shear strain ranges at the risk point in one thermal shock cycle were imported into Equation (2), and the relationship equations of TBC thickness and strain under the *i*th operation condition were fitted as

$$\begin{cases} \Delta\gamma_i = 0.00856 - 4.649 \cdot \delta + 1613 \cdot \delta^2 - 183023 \cdot \delta^3 + 7.58 \times 10^6 \cdot \delta^4 \\ \Delta\varepsilon_i = 0.0058 - 4.213 \cdot \delta + 1308 \cdot \delta^2 - 160674 \cdot \delta^3 + 6.71 \times 10^6 \cdot \delta^4 \end{cases} \tag{12}$$

According to Equation (11), Equation (1) and the cyclic data in Table 5, the coefficients of Equation (2) can be obtained (i.e., *a* = 54, *b* = 3.45). In the same condition, more thermal fatigue test data are provided by Geng [37], and applied to verify the accuracy of the TFL model. The verification results are illustrated in Figure 8. As revealed in Figure 8, all of the TFL points were almost distributed in triple dispersion zone, which supports the validity and feasibility of the fitted phenomenological life prediction model.

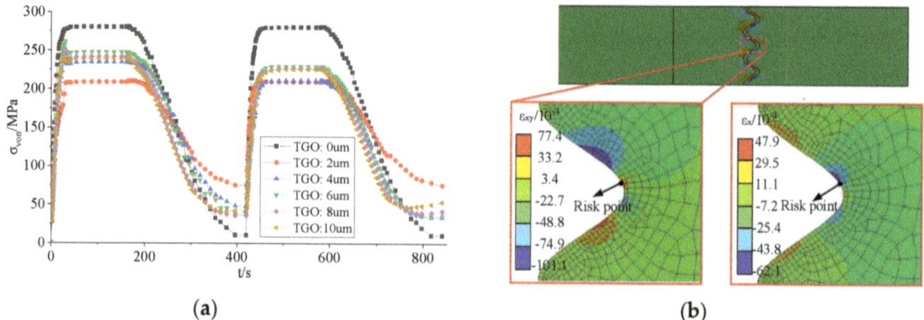

Figure 7. Simulation results of the top coat (TC) layer in ceramic metal tube. (**a**) Maximum von-Mises stress in TC layer; (**b**) Strain contour of 2 μm.

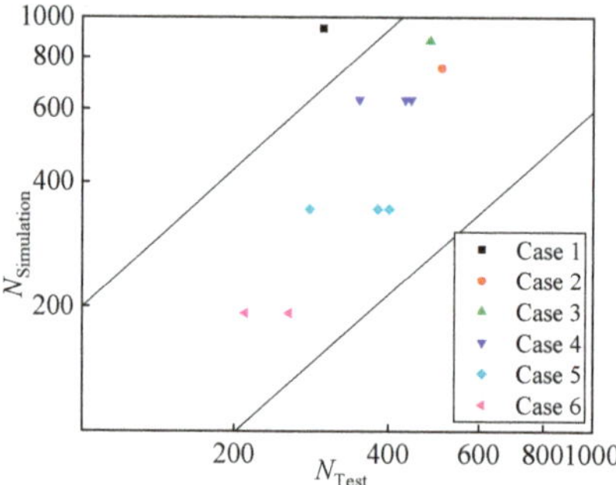

Figure 8. Thermal fatigue test and simulated data [14,37].

4. Thermal Cycle Analysis and Life Prediction of NGV.

4.1. Temperature Field Analysis

In general, the temperature curve of the thermal shock test is a series of trapezoidal waves, thus it is difficult to simulate the heat transfer process considering the time-varying momentum and temperature. As the heat transfer rate of gas is much larger than that of metals [18], the heating process of hot gas was neglected in the numerical simulation. With respect to the simulation study, a typical gas cycle profile is shown in Figure 9.

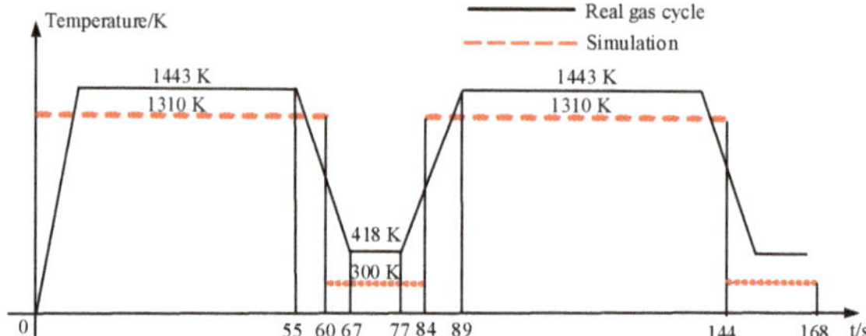

Figure 9. Typical gas cycle profile of the thermal fatigue test [38].

According to the heating cycle, the inlet temperatures of the heating stage and cooling stage were 1310 K and 300 K, respectively. Based on the steady state conjugate heat transfer simulation, the heat transfer boundary conditions were obtained. The convection heat transfer coefficient is:

$$h = \frac{q}{T_{sp} - T_{aw}} \tag{13}$$

where q is the wall heat flux; T_{aw} is the wall temperature; and T_{sp} is the mean temperature at the inlet.

Through different combinations of wall temperature and mean temperature as listed in Table 6, the convection heat transfer coefficients of different domains and stages were calculated. The convective heat transfer coefficient of the heating and cooling stages are shown in Figure 10.

Table 6. Temperature parameters.

Domain	Fatigue Test Cycle	T_{aw}	T_{sp}
Hot gas	Heating stage	300 K	1310 K
	Cooling stage	1000 K	300 K
Cooling air	Heating stage	300 K	500 K
	Cooling stage	1000 K	500 K

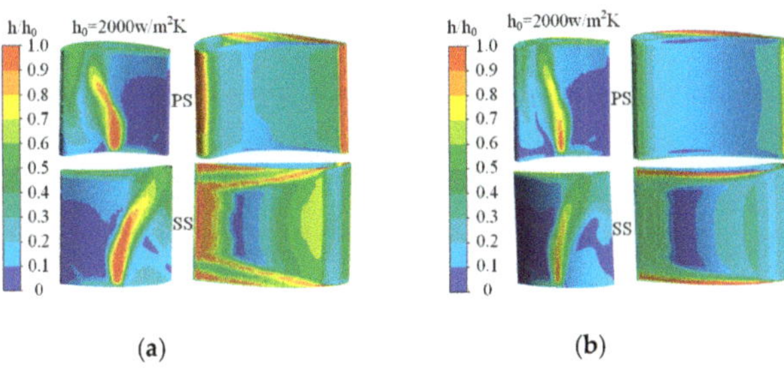

Figure 10. Convection heat transfer coefficient of the heating and cooling stages. (a) Heating stage; (b) Cooling stage.

As revealed in Figure 10, the high-speed heat transfer regions were located on the leading edge, trailing edge, and the center of the inner surface of the master model, respectively. The convection heat transfer coefficient at the leading edge was larger because the leading edge was directed against the flow direction of the hot gas, and the kinetic energy and internal energy were directly converted to surface temperature. At the trailing edge, the high-temperature gas flowed along the surfaces of both the suction side (SS) and pressure side (PS), then mixed to form a strong turbulence intensity. The increase in the turbulent energy increased the heat transfer rate. In addition, the convective heat transfer coefficient of the heating stage was larger than that of the cooling stage due to the high gas speed in a high-temperature environment.

Based on the acquired convective heat transfer coefficient of the heating and cooling stages, the master model was applied to the thermo-elastic coupling calculation of the NGV for 168 s under the thermal cycle, where the heat fluxes at the upper and lower surface were 10,000 w/m^2, referring to the solid rim [39]. The variation trend of the surface temperature field is shown in Figure 11.

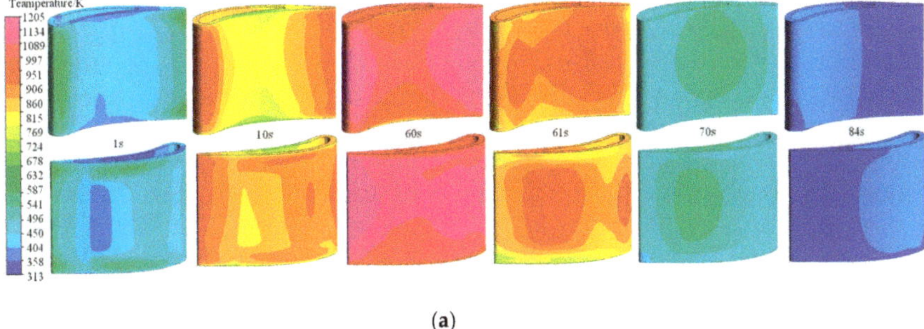

(a)

Figure 11. *Cont.*

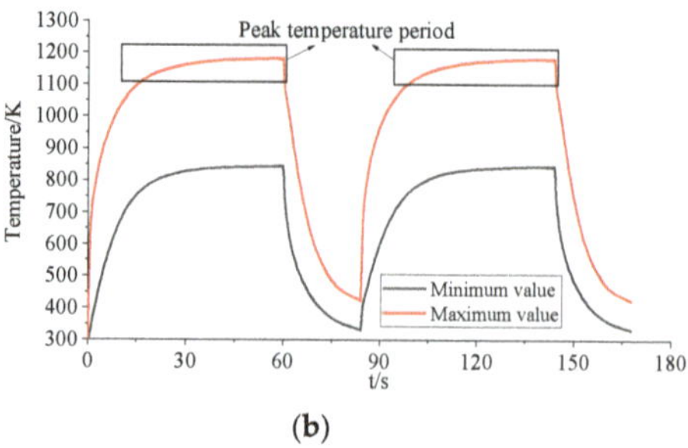

(b)

Figure 11. Variation trend and temperature curves with heating time. (**a**) Change of temperature contour with heating time; (**b**) Changing curves of maximum and minimum temperature.

As shown in Figure 11, the surfaces of the leading edge and trailing edge were heated first, and then reached the maximum temperature quickly. The two areas corresponded to regions A and B in the master model in Figure 4a. In the heat transfer process in Figure 11b, the surface temperature of the NGV rose rapidly toward a stable value after 15 s.

4.2. Temperature Test

To verify the thermal-elastic simulation of the NGV, temperature-sensitive paint (TSP) technology was adopted to measure the surface temperatures of the NGV under the maximum operation conditions of a turbojet engine. The surface temperature tests for the NGV were performed inside an indoor aero-engine test rig at the China Gas Turbine Establishment, the structure of which is shown in Figure 12. The technical parameters of the engine are listed in Table 3.

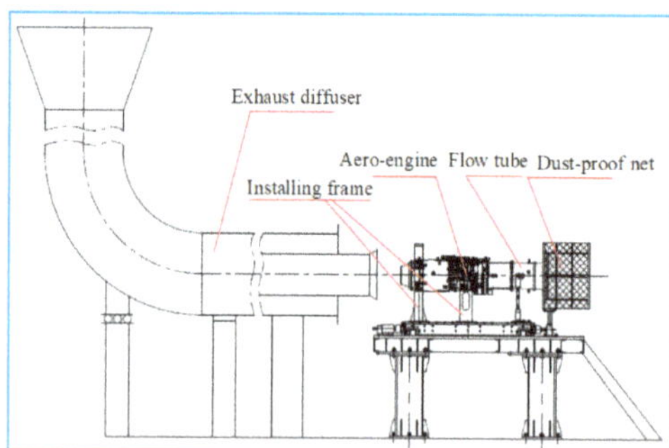

Figure 12. Indoor aeroengine test rig [26].

As each TSP material has one dedicated temperature sensitive range, only temperatures can be recorded in this range in the TSP test. Two typical TSP were selected to show the temperature distributions in the NGV. Isothermal lines (also called discoloration lines) were drawn to calibrate and explain the TSPs instead of the colors after heating. The calibrated temperatures of the thermal-indicate standard models under the constant peak temperature for 180 s are shown in Figure 13. In the coming work, the colors of the TSP-M02 and TSP-M05 were referred to determine the temperature distributions gained in the thermal-elastic simulation of the NGV in Section 4.

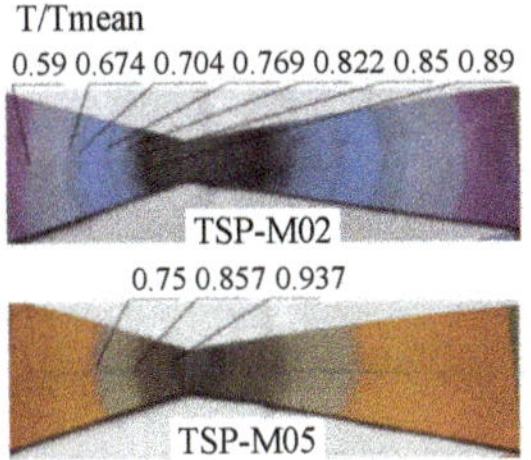

Figure 13. Temperature test of the standard temperature-sensitive paint (TSP) model.

For the high-integrity approach with the master–slave model, the accuracy of the TFL depends highly on the prediction precision of the temperature field. According to the simulation, we found that the temperature distribution of the NGV was stable after 60 s, at which the temperature fields were compared with the result of the TSP test as shown in Figure 14.

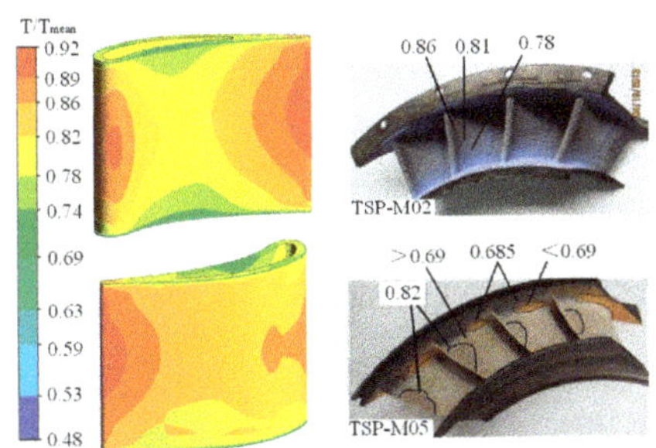

Figure 14. Comparison of the predicted temperature contour with the TSP test.

As revealed in Figure 14, the temperature range of the leading edge and trailing edge were 1021 K to 1126 K and larger than 1070 K, respectively. The maximum temperature of the TSP test was a bit higher than the simulation. The film cooling technology was used to reduce the surface temperature. Therefore, the simulation temperature was reasonable. Compared with traditional methods, the prediction accuracy of the temperature field was improved by using the fully coupled technology in the heat transfer simulation [18]. Furthermore, the temperatures at the center of the leading edge and the trailing edge were 1177 K and 1205 K, respectively, which were similar to that of the ceramic metal tube test and indicate that the parameters in the oxidation weight gain model are also applicable to the NGV.

4.3. Matching of Master Model and Slave Model

To transfer the boundary displacement and body temperature data from the master model (i.e., source) to the slave model (i.e., target), we reasonably adopted the volume element intersection mapping algorithm [24]. The first step in the process of the volume element intersection mapping algorithm is to divide the mapping source mesh into an imaginary structured grid, with each grid section called a "bucket." Next, each node on the data transfer regions of the target mesh is initially associated with a bucket. For the master–slave model discussed in this paper, the subdomain of master model was large enough to hold the mesh element of slave model as shown in Figure 4a. Therefore, the volume element intersection mapping algorithm was used to match each of the target nodes to one source element in the bucket. This was done by looping through all the source elements in that bucket and checking to see whether the target node was within their domain or not. The transferring results are displayed in Figure 15. Obviously, the imported data in the slave model still retained the distribution of temperature and deformation in the master model, although the mesh number of the master model and slave model was different.

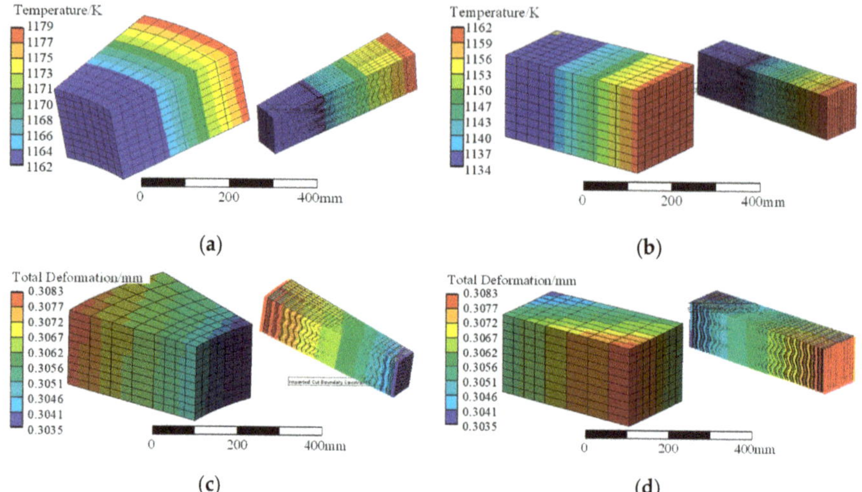

Figure 15. Comparison of parameters before and after importation. (**a**) Temperature in region A; (**b**) Temperature in region B; (**c**) Total deformation in region A; (**d**) Total deformation in region B.

To validate the validity of the volume element intersection mapping algorithm, we performed a high-integrity thermal fatigue life analysis with the transferred slave model without both TGO and sinusoidal interface. The thermal stress comparison on the TC layer of master model and slave model is shown in Figure 16. The results in Figure 16 show that in the same thermal cycle, the stress variations of the slave model agreed well with the master model, because the maximum error was less than 1%. Furthermore, the trailing edge endured a large thermal shock load, especially at the beginning of the heating and cooling stage. Compared to Figure 11, the heat transfer rate at the trailing edge was much larger than that of the leading edge, so that under a high thermal load, a large temperature gradient was produced to induce a high thermal stress level. Therefore, compared with the traditional simulation method, the high-integrity analysis approach is more likely to find out the essential reasons that lead to thermal fatigue failure.

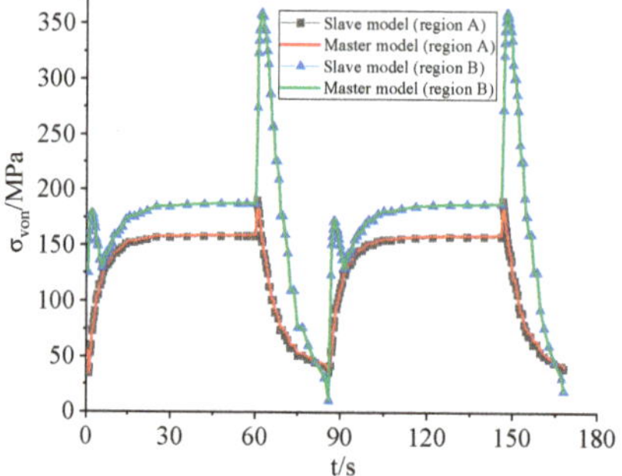

Figure 16. Comparison of maximum thermal stress in the TC layer by different models in regions A and B.

4.4. Thermal Fatigue Life Analysis of Nozzle Guide Vane with TBC

When the TGO thickness was 2 μm, the thermal stress simulation results in the TC layer based on the slave model with sinusoidal interface are as shown in Figure 17. Compared to Figure 16, the thermal stress in Figure 17 changed smoothly on the sinusoidal interface, because the shape of the mixed TC/TGO layer was a buffer. Furthermore, we found that the risk moment was at the beginning of the cooling stage. The strain contours of the slave model at 60.6 s are drawn in Figure 18.

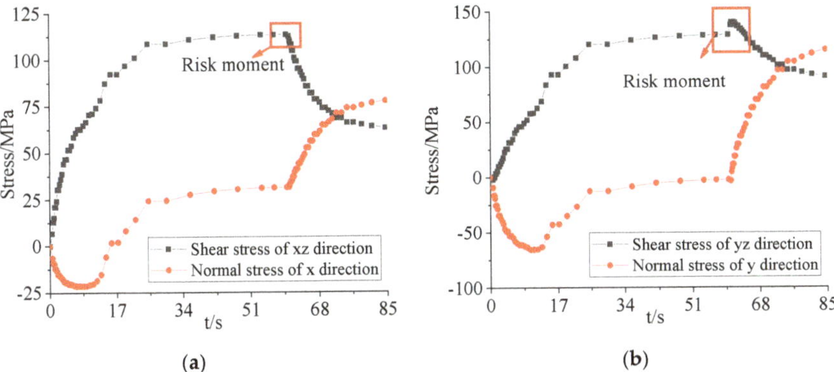

Figure 17. Stress curves in the slave model. (**a**) TC layer at the leading edge; (**b**) TC layer at the trailing edge.

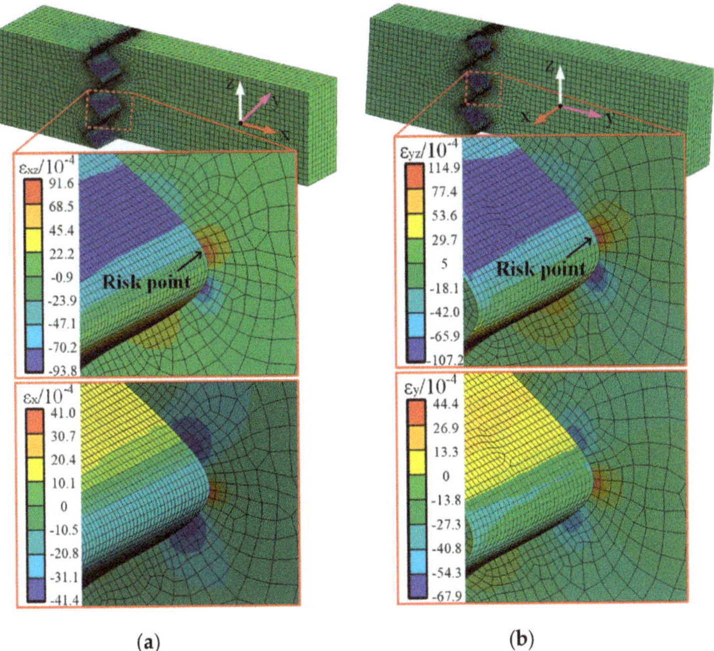

Figure 18. Strain contours of the slave model. (**a**) Leading edge; (**b**) Trailing edge.

As shown in Figure 18, the peaks of the shear strain and axial strain occurred near the mixed TC/TGO layer, and the maximum strain point was in the valley region. The distributions of the shear and axial strain were basically consistent with those of the ceramic tube in Figure 7b. The reason for this is that the thermal stress level of the mixed TC/TGO layer at the trailing edge was affected by the total deformation of the master model. The leading edge and trailing edge can be considered as a series of ceramic tubes with different radii, which is also the reason for why the distribution of the thermal strain was similar to the slave model and ceramic tube model. The stress and strain at the risk point in the mixed TC/TGO layer were almost independent with the radius of the ceramic tube [14]. As a result, the TFL model of ceramic tubes previously established is also applicable to the high-integrity life prediction of a NGV. The grid node at the shear strain concentration points highlighted in Figure 18 was regarded as the crack initiation point (i.e., risk point) as the shear strain was much higher than the axial strain. The stress–strain curves at the risk point are shown in Figure 19.

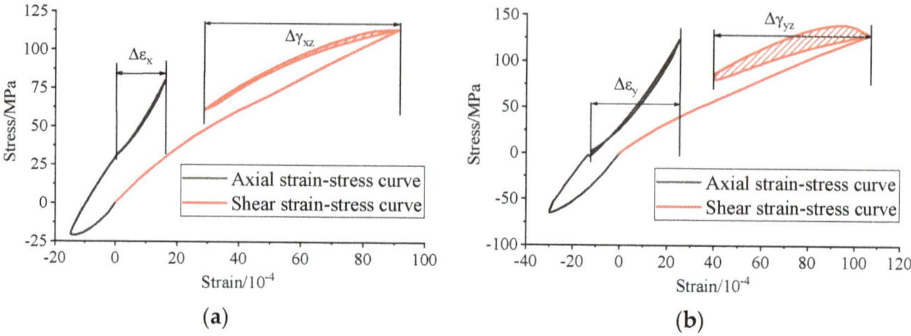

Figure 19. Stress–strain curve at the risk point of the slave model. (**a**) Leading edge; (**b**) Trailing edge.

As revealed in Figure 19, the integral area (red and black region) and strain range of the stress–strain curve of the trailing edge were larger than that of the leading edge (black region), indicating that the thermal shock attacking at the leading edge has a stronger failure energy under the same thermal cycle. On the other side, the fatigue hysteresis loops at the risk point basically reached a stable state, indicating that the thermal stress of the TC layer reached the plastic stability stage after two thermal cycles (i.e., the axis and shear strain range do not vary by repeating the thermal cycle simulation). Thus, the thermal strain range at the second thermal cycle was inputted into the TFL model.

By repeatedly producing the high-integrity TFL analysis of the NGV, the thermal strain ranges of six TGO thicknesses (i.e., 0, 2 μm, 4 μm, 6 μm, 8 μm, and 10 μm) were obtained as shown in Figure 20.

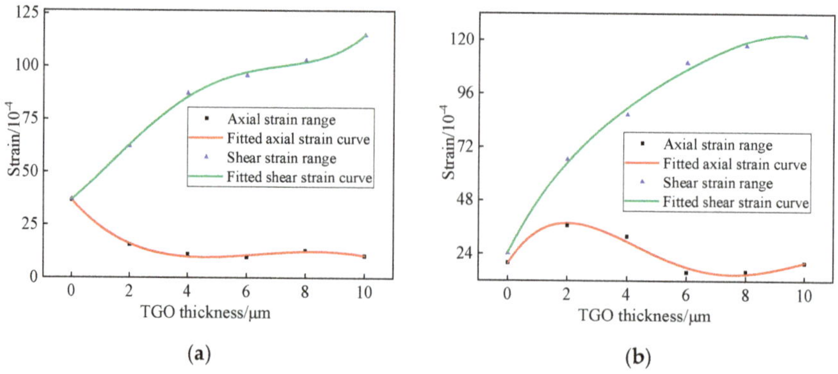

Figure 20. Strain ranges and fitted curves. (**a**) Region A; (**b**) Region B.

As shown in Figure 20, the strain levels at the sinusoidal interface regions were correlated with TGO thickness. The axial strain ranges were basically smaller than the shear strain range, and the changing trends of the axial strain ranges at both the leading edge and trailing edge were opposite the shear strain ranges. The strain ranges of the TC layer did not change much at the beginning of the thermal cycles, but the shear strain ranged increased obviously with the increase in the TGO thickness, which indicates that shear strain is a key factor inducing the thermal fatigue of the TBC material. The varying trends of strain range at the leading and trailing edge were similar, proving that the corner radius slightly influences the micro-strain. The fitted equations of the TGO thickness and strain range at the risk points are as follows.

Strain range of leading edge:

$$\begin{cases} \Delta\gamma_i = 0.00369 + 1.139 \cdot \delta + 172 \cdot \delta^2 - 49908 \cdot \delta^3 + 2.91 \times 10^6 \cdot \delta^4 \\ \Delta\varepsilon_i = 0.00364 - 1.508 \cdot \delta + 283 \cdot \delta^2 - 1933 \cdot \delta^3 + 3.49 \times 10^5 \cdot \delta^4 \end{cases} \tag{14}$$

Strain range of trailing edge:

$$\begin{cases} \Delta\gamma_i = 0.0025 + 2.64 \cdot \delta - 380 \cdot \delta^2 + 38204 \cdot \delta^3 - 1.69 \times 10^6 \cdot \delta^4 \\ \Delta\varepsilon_i = 0.00198 + 2.1 \cdot \delta - 753 \cdot \delta^2 + 84741 \cdot \delta^3 - 3.04 \times 10^6 \cdot \delta^4 \end{cases} \tag{15}$$

Importing Equations (13) and (14) into the TFL model, the cumulative damage is calculated to gain the number of the TBC cyclic life as shown in Table 7.

Table 7. Thermal fatigue life of the TBC on the NGV.

Number	Life Cycle		Thermal Fatigue Life	
	Leading Edge, N	Trailing Edge, N	Leading Edge, H	Trailing Edge, H
Case 1	1892	1558	40	33
Case 2	1345	870	279	180
Case 3	1419	916	30	20
Case 4	1310	704	28	15
Case 5	1126	558	24	12
Case 6	895	536	19	11

As demonstrated in Table 7, the service life of the NGV was reduced with the increase in the TGO thickness. The thermal life of the TBC at the trailing edge was shorter than that at the leading edge. Darolia [5] found that the peel off life cycle of the TBC on a NGV with a NiAl composition at 1448 K was about 1500 times. The thermal cyclic test in [40] clarified that 5–10% spallation of the BC layer happened during 1500–2000 cycles under the temperature of 1323 K. Therefore, the developed master–slave model was validated to be sufficiently accurate for the high-integrity life prediction.

5. Conclusions

The target of this paper was to propose an efficient master–slave model for the thermal fatigue life (TFL) prediction method of a nozzle guide vane (NGV) with a thermal barrier coat (TBC), with respect to flow-thermo-structural coupling, oxidation damage, and thermal fatigue damage. The modeling and simulation methods and life molding processes were first investigated and validated. Then, the TFL prediction method of the NGV with the TBC was performed with respect to the proposed master–slave model in the foundation of the temperature filed analysis, temperature test, and matching of the master and slave models. Through these studies, some conclusions and findings can be summarized as follows:

(1) The adopted SST γ-θ turbulence model effectively predicted the convective heat transfer coefficient by thermal-elastic simulation, and the temperature field solved by the fully coupled solid 226 element agreed well with the test data of the aircraft engine. This revealed that the fully coupling solid 226 element is promising to improve the accuracy of the temperature field calculation. Compared with the traditional method, the accuracy of the temperature field calculation was improved by over 5%.

(2) Based on the volume element intersection mapping algorithm, the boundary conditions of the temperature and displacement were successfully transferred from the master model into the slave model with high precision. The simulation results of the master model and slave model showed good agreement, indicating that it is possible to link macro- and micro-scales with the master–slave model presented in this paper.

(3) The ranges of the axial and shear strains in the TC layer were affected by the thickness of the TGO layer, but the trends of the two kind strains were almost the opposite. With the increase in the thermal cyclic number, shear strain plays a dominant role in the life model because it seriously effects the TFL prediction of the TBC on a NGV for an aero engine.

(4) Based on the developed master–slave model, the TFL of a typical NGV was precisely predicted with minor errors when compared to the test data, and the life variation also met the actual usage of the NGV. The proposed method comprehensively considers the physical characteristics of heat transfer boundaries, macro-scale and micro-scale to reflect the coupling failure of oxidation damage and thermal fatigue.

In short, the developed master–slave model was validated to be highly-computationally precise and efficient with regard to TFL prediction for a NGV with a TBC by comparing the temperature and the life cycle of the key points at the leading edge and trailing edge with the experiments. The efforts of the study provide a promising modeling strategy (master–slave model) for the integrated TFL prediction design of thermal structures other than NGV in engineering.

Author Contributions: Conceptualization, Y.A. and P.G.; Methodology, P.G. and Y.Y.; Software, P.G. and Y.Y.; Validation, P.G. and C.F.; Formal Analysis, P.G. and C.F.; Investigation, P.G. and C.F.; Resources, Y.A. and C.F.; Data Curation, P.G. and C.F.; Writing-Original Draft Preparation, P.G.; Writing-Review & Editing, C.F.; Visualization, P.G.; Supervision, Y.A. and C.F.; Project Administration, Y.A. and C.F.; Funding Acquisition, Y.A. and C.F.

Acknowledgments: This work was sponsored by the National Natural Science Foundation of China (Grant nos. 51975124 and 51605016), and the Research Start-Up Funding of the Fudan University (Grant no. FDU38341). All authors would like to thank them for their support.

Conflicts of Interest: The authors declare that there are no conflicts of interest regarding the publication of this article.

Nomenclature

P	pressure (Pa)
T	temperature (K)
t	time (s)
h	convective heat transfer coefficient (W/m·K)
C	specific heat capacity (J/kg·K)
E	Young's modulus (Pa)
K	heat transfer coefficient (W/m·K)
W	air mass flow (kg/s)
Ma	Mach number
w	percentage of cold coolant flow (%)
F	gas-oil ratio
i	number of temperature point
q	wall heat flux
Q	constant
n	constant
a	constant
b	constant
c	constant
n_m	cycle number
N	cyclic Life

Greek letters

ν	Poisson's ratio
σ	stress (MPa)
π	compressor pressure ratio
δ	thickness of thermal barrier coat (μm)
γ	shear strain
ε	axial strain
σ_y	yield stress (Pa)

Subscript

von	von Mises
eng	engine inlet
com	compressor
ex	exhaust nozzle
*	total
c	coolant
fu	fuel
sh	shaft
ca	cooling air inlet
mean	average temperature
g	ideal gas
rs	combustor
in	mainstream inlet
out	mainstream outlet
aw	adiabatic wall
sp	specified

Acronyms

NGV	nozzle guide vane
SS	section side
PS	pressure side
TBC	thermal barrier coat
TC	top coat
BC	bond coat
TGO	thermally grown oxide
CFD	computational fluid dynamics
FTS	flow-thermo-structural
FV	finite volume
FE	finite element
HTC	heat transfer coefficient
TFL	thermal fatigue life
TSP	temperature-sensitive paint

References

1. Ebrahimi, H.; Nakhodchi, S. Thermal fatigue testing and simulation of an APS TBC system in presence of a constant bending load. *Int. J. Fatigue* **2017**, *96*, 1–9. [CrossRef]
2. Zhang, C.Y.; Wei, J.S.; Jing, H.Z.; Fei, C.W.; Tang, W.Z. Reliability-Based Low Fatigue Life Analysis of Turbine Blisk with Generalized Regression Extreme Neural Network Method. *Materials* **2019**, *12*, 1545. [CrossRef] [PubMed]
3. Zhang, C.Y.; Yuan, Z.S.; Wang, Z.; Fei, C.W.; Lu, C. Probabilistic Fatigue/Creep Optimization of Turbine Bladed Disk with Fuzzy Multi-Extremum Response Surface Method. *Materials* **2019**, *12*, 3367.
4. Sheffler, K.D.; Gupta, D.K. Current status and future trends in turbine application of thermal barrier. *ASME J. Eng. Gas Turbine Power* **1988**, *110*, 605–609. [CrossRef]
5. Darolia, R. Thermal barrier coatings technology: Critical review, progress update, remaining challenges and prospects. *Int. Mater. Rev.* **2013**, *58*, 315–348. [CrossRef]

6. Ranjbar-Far, M.; Absi, J.; Mariaux, G.; Dubois, F. Simulation of the effect of material properties and interface roughness on the stress distribution in thermal barrier coatings using finite element method. *Mater. Des.* **2010**, *31*, 772–781. [CrossRef]

7. Robert, E.; Kang, Y.; Sten, J.; Ru, L.P. Microstructure-Based Life Prediction of Thermal Barrier Coatings. *Key Eng. Mater.* **2013**, *592–593*, 413–416.

8. Song, H.; Lee, J.M.; Kim, Y.; Oh, C.S. Evaluation of Effect on Thermal Fatigue Life Considering TGO Growth. *J. Korean Soc. Prec. Eng.* **2014**, *31*, 1155–1159. [CrossRef]

9. Guan, P.; Ai, Y.T.; Fei, C.W. An enhanced flow-thermo-structural modeling and validation for the integrated analysis of film cooling nozzle guide vane. *Energies* **2019**, *12*, 2775. [CrossRef]

10. Xie, W.G.; Eric, J.; Maurice, G. Stress and cracking behavior of plasma sprayed thermal barrier coatings using an advanced constitutive model. *Mater. Sci. Eng. A* **2006**, *419*, 50–58. [CrossRef]

11. Song, L.K.; Bai, G.C.; Fei, C.W.; Tang, W.Z. Multi-failure probabilistic design for turbine bladed disks using neural network regression with distributed collaborative strategy. *Aerosp. Sci. Technol.* **2019**, *92*, 464–477. [CrossRef]

12. Yang, X.G.; Geng, Y.; Zhou, H.P. An Experimental Study of Oxidation and Thermal Fatigue of TBC. *J. Aerosp. Power* **2003**, *18*, 195–200.

13. Yang, X.G.; Geng, Y.; Zhou, H.P. A Study of Thermal Fatigue Life Prediction of TBC. *J. Aerosp. Power* **2003**, *18*, 201–205.

14. Wei, H.L.; Yang, X.G.; Qi, H.Y. Study on thermal fatigue life prediction for plasma sprayed thermal barrier coatings on the surface of turbine vane. *J. Aerosp. Power* **2008**, *23*, 1–8.

15. Kim, K.M.; Park, J.S.; Dong, H.L.; Lee, T.W.; Cho, H.H. Analysis of conjugated heat transfer, stress and failure in a gas turbine blade with circular cooling passages. *Eng. Fail. Anal.* **2011**, *18*, 1212–1222. [CrossRef]

16. Kim, K.M.; Yun, H.J.; Yun, N.; Dong, H.L.; Cho, H.H. Thermo-mechanical life prediction for material lifetime improvement of an internal cooling system in a combustion liner. *Energy* **2011**, *36*, 942–949. [CrossRef]

17. Chung, H.; Sohn, H.S.; Park, J.S.; Kim, K.M.; Cho, H.H. Thermo-structural analysis of cracks on gas turbine vane segment having multiple airfoils. *Energy* **2017**, *118*, 1275–1285. [CrossRef]

18. Guan, P.; Ai, Y.T.; Wang, Z.; Wang, F. Numerical Simulation of Nozzle Guide Vane Subjected to Thermal Shock Load. *J. Propuls. Technol.* **2016**, *37*, 1938–1945.

19. Zhu, W.; Wang, J.W.; Yanga, L.; Zhoua, Y.C.; Wei, Y.G.; Wu, R.T. Modeling and simulation of the temperature and stress fields in a 3D turbine blade coated with thermal barrier coatings. *Surf. Coat. Technol.* **2017**, *315*, 443–453. [CrossRef]

20. Staroselsky, A.; Martin, T.J.; Cassenti, B. Transient thermal analysis and viscoplastic damage model for life prediction of turbine components. *J. Eng. Gas Turbines Power* **2015**, *137*, 042501. [CrossRef]

21. Johnson, J.J.; King, P.I.; Clark, J.P.; Koch, P.J. Exploring conjugate CFD heat transfer characteristics for a film-cooled flat plate and 3-D turbine inlet vane. In Proceedings of the ASME Turbo Expo, Copenhagen, Denmark, 11–15 June 2012.

22. Meier, S.M.; Nissley, D.M.; Sheffler, K.D. *Thermal Barrier Coating Life Prediction Model Development, Phase II Final Report*; NASA-CR-189111; Pratt and Whitney Aircraft: East Hartford, CT, USA, 1991.

23. China Aviation Materials Manual Commission. *China Aeronautical Materials Handbook*; Standards Press of China: Beijing, China, 2002; Volume 9, pp. 117–128.

24. Barnett, R.M.; Murphy, M.; Deutsch, C.V. A grid mapping algorithm for modeling with geometric transforms. *Comp. Geosci.* **2019**, *126*, 9–20. [CrossRef]

25. Nowinski, J.L. *Theory of Thermoelasticity with Applications*; Martinus Nijhoff Publishers: Hague, The Netherlands, 1978.

26. Xiong, Q.R.; Shi, X.J.; Xu, F.; Zhong, M. Surface temperature measurement of turbine nozzle based on temperature-sensitive paint. *Gas Turbine Exp. Res.* **2014**, *27*, 45–48.

27. Zhu, W.; Cai, M.; Yanga, L.; Guod, J.W.; Zhoua, Y.C.; Lue, C. The effect of morphology of thermally grown oxide on the stress field in a turbine blade with thermal barrier coatings. *Surf. Coat. Technol.* **2015**, *276*, 160–167. [CrossRef]

28. Bao, T.N. Numerical Study on Thermal Shock of Turbine Guide Vane Considering Thermoelasticity Coupling Effect. Ph.D. Thesis, Shenyang Aerospace University, Shenyang, China, 2018; pp. 18–26.

29. Kim, G.M.; Yanar, N.M.; Hewitt, E.N.; Pettit, F.S.; Meier, G.H. The effect of the type of thermal exposure on the durability of thermal barrier coatings. *Scr. Mater.* **2002**, *46*, 489–495. [CrossRef]

30. Rösler, J.; Bäker, M.; Aufzug, K. A parametric study of the stress state of thermal barrier coatings: Part I: Creep relaxation. *Acta Mater.* **2004**, *52*, 4809–4817.

31. Veal, B.W.; Paulikas, A.P.; Hou, P.Y. Tensile stress and creep in thermally grown oxide. *Nat. Mater.* **2006**, *5*, 349–351. [CrossRef]

32. Wang, X.; Xiao, P. Residual stresses and constrained sintering of YSZ/Al2O3 composite coatings. *Acta Mater.* **2004**, *52*, 2591–2603. [CrossRef]

33. Fei, C.W.; Lu, C.; Liem, P.R. Decomposed-coordinated surrogate modeling strategy for compound function approximation in a turbine-blisk reliability evaluation. *Aerosp. Sci. Technol.* **2019**, 105466. [CrossRef]

34. Lian, X.C.; Wu, H. *Principles of Aeroengine*; Northwest Polytechnic University Press: Xi'an, China, 2005; pp. 152–161.

35. Wang, Z.; Wang, D.; Liu, Z.; Feng, Z. Numerical analysis on effects of inlet pressure and temperature non-uniformities on aero-thermal performance of a HP turbine. *Int. J. Heat Mass Transf.* **2017**, *104*, 83–97. [CrossRef]

36. Wang, Z.D.; Liu, Z.F.; Feng, Z.P. Influence of mainstream turbulence intensity on heat transfer characteristics of a high pressure turbine stage with inlet hot streak. *ASME J. Turbomach.* **2015**, *138*, 041005. [CrossRef]

37. Geng, D. Strength Analysis and Life Prediction Study on Thermal Barrier Coatings. Ph.D. Thesis, Beijing University of Aeronautics and Astronautics, Beijing, China, 1990.

38. Peng, Z.Y.; Lu, W.L. A comparative test approach for thermal fatigue life of high-heat blade under thermal shock. *J. Aerosp. Power* **1997**, *12*, 33–36.

39. Guan, P.; Ai, Y.T.; Wang, Z.; Bao, T.N. Structure strength simulation of film cooling vane after heat shock by thermal/flow/structure coupling. *J. Aerosp. Power* **2018**, *33*, 1811–1820.

40. Jonnalagadda, K.P.; Eriksson, R.; Lib, X.H.; Peng, R.L. Thermal barrier coatings: Life model development and validation. *Surf. Coat. Technol.* **2019**, *362*, 293–301. [CrossRef]

© 2019 by the authors. Licensee MDPI, Basel, Switzerland. This article is an open access article distributed under the terms and conditions of the Creative Commons Attribution (CC BY) license (http://creativecommons.org/licenses/by/4.0/).

MDPI
St. Alban-Anlage 66
4052 Basel
Switzerland
Tel. +41 61 683 77 34
Fax +41 61 302 89 18
www.mdpi.com

Applied Sciences Editorial Office
E-mail: applsci@mdpi.com
www.mdpi.com/journal/applsci

www.ingramcontent.com/pod-product-compliance
Lightning Source LLC
LaVergne TN
LVHW070920170726
843515LV00005B/829